PENGUIN

DISCOURSE ON METHOD AND RELATED WRITINGS

RENÉ DESCARTES was born in 1596 at La Haye (now called Descartes) near Tours, and educated at the Jesuit College at La Flèche. Like many of his contemporaries he contested the value of an education based on Aristotelianism and, after leaving college, attempted to resolve the sceptical crisis of his age by devising a method of reasoning modelled on the rigour and certainty of mathematics. Despite claiming to avoid theological questions and stay within the scope of human reason, his writings involved him in numerous disputes with theologians of both the Catholic and (especially) the Reformed persuasion. In 1621, after a period spent in the Netherlands, Bohemia and Hungary as a soldier, he left the army and devoted himself to the study of science and philosophy. He retired to the Netherlands in 1628 and spent the next twenty-one years there, living and working in seclusion. Then in late 1649, after an invitation the previous year, he went to Sweden to take up a post instructing Queen Christina in philosophy. Descartes's habit for many years was to rise not much before midmorning but the Queen wished to be tutored at five o'clock in the morning, three days a week. This and the cold weather placed a severe strain on Descartes's health and he contracted pneumonia, dying in Stockholm in early 1650, having only just begun to teach the Queen. His last words were reportedly *a mon âme, il faut partir* ('my soul, it is time to depart').

DESMOND CLARKE was educated in Ireland, France and the United States. He is Professor of Philosophy at the National University of Ireland, Cork, and a member of the Royal Irish Academy. His publications include *Descartes' Philosophy of Science* (1982), *Church & State* (1984), *Descartes's Theory of Mind* (Oxford, 2003) and two translations of early Cartesian authors: Poulain de la Barre's *The Equality of the Sexes* (1990) and Louis de la Forge's *Treatise on the Human Mind* (1997). He is co-editor with Karl Ameriks of the series Cambridge Texts in the History of Philosophy. He is married, with two daughters, Nora and Ann.

The new Penguin classics edition of Descartes consists of this volume and its companion volume of writings, *Meditations and Other Metaphysical Writings*.

RENÉ DESCARTES

Discourse on Method
and Related Writings

Translated with an Introduction by
DESMOND M. CLARKE

PENGUIN BOOKS

For Ann

PENGUIN BOOKS

Published by the Penguin Group
Penguin Books Ltd, 80 Strand, London WC2R ORL, England
Penguin Putnam Inc., 375 Hudson Street, New York, New York 10014, USA
Penguin Books Australia Ltd, 250 Camberwell Road, Camberwell, Victoria 3124, Australia
Penguin Books Canada Ltd, 10 Alcorn Avenue, Toronto, Ontario, Canada M4V 3B2
Penguin Books India (P) Ltd, 11 Community Centre, Panchsheel Park, New Delhi – 110 017, India
Penguin Books (NZ) Ltd, Cnr Rosedale and Airborne Roads, Albany, Auckland, New Zealand
Penguin Books (South Africa) (Pty) Ltd, 24 Sturdee Avenue, Rosebank 2196, South Africa

Penguin Books Ltd, Registered Offices: 80 Strand, London WC2R ORL, England

www.penguin.com

First published 1637
This translation first published by Penguin Books 1999
Reprinted with a new Chronology and updated Further Reading 2003

040

Translation, introduction and editorial matter copyright © Desmond M. Clarke, 1999, 2003
All rights reserved

The moral right of the translator and editor has been asserted

Set in 10.25/12.5pt Monotype Van Dijck
Typeset by Rowland Phototypesetting Ltd,
Bury St Edmunds, Suffolk
Printed and bound in Great Britain by Clays Ltd, Elcograf S.p.A.

Except in the United States of America, this book is sold subject
to the condition that it shall not, by way of trade or otherwise, be lent,
re-sold, hired out, or otherwise circulated without the publisher's
prior consent in any form of binding or cover other than that in
which it is published and without a similar condition including this
condition being imposed on the subsequent purchaser

ISBN-13: 978-0-140-44699-9

www.greenpenguin.co.uk

Penguin Books is committed to a sustainable
future for our business, our readers and our planet.
This book is made from Forest Stewardship
Council™ certified paper.

Contents

ACKNOWLEDGEMENTS	vi
NOTE ON REFERENCES TO DESCARTES	vii
CHRONOLOGY	ix
INTRODUCTION	xi
FURTHER READING	xxxv
Discourse on the Method for Guiding One's Reason and Searching for Truth in the Sciences	1
Selected Correspondence, 1636–9	55
The World, or a Treatise on Light (Chapters 1–7)	81
Rules for Guiding One's Intelligence in Searching for the Truth	113
TEXT NOTES	195
INDEX	205

Acknowledgements

I am grateful to the Arts Faculty Research Fund, University College, Cork, for a travel grant which made it possible to consult first editions and early translations of Cartesian texts at the British Library.

Desmond M. Clarke
Cork, 11 October 1998

Note on References to Descartes

The standard edition of Descartes's works was prepared by Charles Adam and Paul Tannery at the beginning of the twentieth century and was published in twelve volumes (L. Cerf, Paris, 1897–1913). These have been revised under the general direction of the Centre National de la Recherche Scientifique, France, and have been reissued as *Oeuvres de Descartes*, ed. C. Adam and P. Tannery (Vrin, Paris, 1964–74). All references in notes to Descartes's works are to this latter edition, and are identified as AT, followed by the volume and page number. References which include only a page number refer to texts translated in this volume. Finally, 'CSM' refers to the three-volume edition of *The Philosophical Writings of Descartes* (see Further Reading).

Chronology

31 March 1596	Born at La Haye (since renamed Descartes), France
16 May 1597	Descartes' mother died
1607–1615	Attended Jesuit College of La Flèche
14 May 1610	Henry IV assassinated
9/10 November 1616	Graduated in Law, University of Poitiers
1618–19	At Breda, in the United Provinces
1618–48	Thirty Years War
1619–21	Travels to Poland and Germany
10 November 1619	Dreams of a 'great discovery' at Neuberg (Bavaria)
1620	Bacon publishes the *New Organon*
March 1621	Hugo Grotius escapes from prison and goes into exile in Paris
1623–8	Travels to Italy, and resides in Paris and Brittany
1627–8	Siege of La Rochelle
1628	Abandons writing the *Rules for Guiding One's Intelligence in Searching for the Truth*
1628–49	Lives at various places in the United Provinces
1633	Galileo's *Dialogue* condemned by the Roman Inquisition
1633	Decides not to publish *The World*
1637	Publishes *Discourse on the Method for Guiding One's Reason, the Dioptrics, the Meteors, and the Geometry*
1639	Campanella dies in exile in Paris, having been imprisoned in Italy for thirty years
1641	Publishes *Meditations on First Philosophy*

CHRONOLOGY

1642	Second edition, expanded and corrected, of the *Meditations*
1644	Publishes *The Principles of Philosophy*
1647	Publishes the French edition of *The Principles of Philosophy*
1648	Publishes response to Regius: *Comments on a Certain Manifesto*
1649	Publishes *The Passions of the Soul*
October 1649	Arrives in Sweden, at the Court of Queen Christina
11 February 1650	Dies at Stockholm
1662	*Treatise on Man* published in Latin
1664	*Treatise on Man* and *The World* published in French
1657, 1659, 1667	Three volumes of Descartes' letters edited by Clerselier
1701	*Rules for Guiding One's Intelligence* published

Introduction

Something radically new occurred in the seventeenth century which made a lasting impression on our way of thinking about the universe in which we live. The philosophical and theological commonplaces of the previous ages were challenged by novel developments in astronomy, physics and biology; the combined impact of these discoveries was so fundamental that it was appropriately called a scientific revolution. This is not to suggest that the change was so sudden that there was no science before 1600 – even in astronomy, two of the foremost contributors to the new world-view, Nicolas Copernicus and Tycho Brahe, published their results in the sixteenth century. None the less, it is helpful to acknowledge the unprecedented scale and significance of scientific developments in the seventeenth century, and to mark their originality accordingly. It was a genuine revolution.

Among those who contributed to the scientific revolution were William Gilbert (magnetism), William Harvey (physiology), Johannes Kepler (astronomy), Galileo Galilei (astronomy and physics), Robert Boyle (chemistry), Christiaan Huygens (optics), Pierre Fermat (mathematics), Thomas Willis (anatomy) and Isaac Newton (physics). These were practising scientists, whose work was discussed both by those who were adequately trained to do so, especially in mathematics, and by a much wider educated readership, to whom the implications of the new sciences were explained. The publication of scientific research and the dissemination of results were organized through a network of societies in Europe which were ranked at different levels to match the interests of members. They included the Royal Society in England and the Académie Royale des Sciences in France, both of which were founded in the 1660s for professional or dedicated scientists, and in which membership was limited; they also included an extensive range of salons

INTRODUCTION

and less professional groups where the results of new scientific and philosophical developments were openly discussed. The widespread interest in scientific theories and their application to practical arts, from waging war to healing the sick, contributed to the development of a new public culture in which the discoveries of experimental scientists were prominently acclaimed.

Of course, scientists of the seventeenth century did not work in a conceptual vacuum; their research institutes were not insulated from theological or philosophical criticism. In parallel with new discoveries in chemistry, physics or astronomy, these scientists reflected on the new methods they used and on the kind of knowledge that resulted from scientific investigations. It was impossible to support a heliocentric system, such as that proposed by Copernicus and Galileo, without being aware of the apparent problems it caused for biblical interpretation; it was equally impossible to speculate about the viability of a mechanistic biology without realizing its implications for traditional accounts of the human mind. Thus scientists of the early modern period were forced to address questions about the scope of different disciplines, the methods used in their development and the contentious issues that arose in apparent or genuine conflicts between the competing claims of traditional learning and the new sciences. In a word, they had to explain and defend the scope and limits of scientific method. It is widely recognized that Descartes made a significant contribution to this task.

René Descartes was born on 31 March 1596 in La Haye (now renamed Descartes), near the city of Tours in France, and was educated at the Jesuit college at La Flèche. As a student there, he followed the standard curriculum of Latin and Greek, mathematics and scholastic philosophy. He was also exposed to contemporary developments in the sciences, or at least to those that were considered relevant to the education of a Catholic gentleman. He left La Flèche in 1615, at the age of nineteen, and enrolled in the Law Faculty at the University of Poitiers, from which he graduated in 1616. Instead of following his father into a legal career, he left France and joined the army of the Prince of Orange, Maurice of Nassau, and later that of Maximilian I. It was during his brief military career that, in 1619, he had the famous dream to which he refers in Part Two of the *Discourse on Method*, and in the course of which he conceived of the project of re-establishing all human knowledge on firm foundations. The path from the initial dream to its eventual realization was a rather complex one, and is discussed in more detail below.

Descartes never held a teaching post at a college or university and, as far as we know, never applied for one. Nor did he earn a living, in the usual sense of that phrase today. Instead, he was able to support himself from personal or family resources, and to devote his time fully to scientific research and philosophical reflection. Beginning in the 1620s, Descartes undertook a number of studies in mathematics, optics and music, and also travelled extensively in Europe. He moved to the Netherlands in about 1628, at the age of thirty-three, and spent the subsequent twenty-one years there doing experiments, writing, and corresponding with other natural philosophers in Europe. While living in the Netherlands, Descartes changed his address frequently, moving from one town to another, mostly to protect his privacy and avoid controversy. During this period all his principal published works appeared: the *Discourse on Method, Dioptrics, Meteors and Geometry* (1637), the *Meditations* (1641) and the *Principles of Philosophy* (1644). In the 1640s, he was involved in on-going philosophical and theological controversies with various Dutch professors, and wrote *Comments on a Certain Manifesto* (1648) as a reply to one of his critics. His final publication, the *Passions of the Soul*, appeared in 1649, the same year in which he accepted an invitation from Queen Christina to provide philosophy tutorials at her court in Stockholm. At the end of August 1649, he travelled to Sweden, and he died there of pneumonia on 11 February 1650.

One might think of the books published during Descartes's lifetime as the official or authoritative sources for understanding his philosophy. However, even his published work raises problems of interpretation. Books written originally in French, such as the essays of 1637, were subsequently published in Latin editions; and those published in Latin, such as the *Meditations* and the *Principles*, were followed by French translations. In many cases, variations between an original text and its translation go far beyond what might be explained by loose translation. Important clarifications and extra sentences were added, while other phrases were deleted. The extent to which Descartes contributed to these changes or endorsed them is not always clear. Thus any significant change may be understood in different ways: it may represent a development in Descartes's thought, it may be a change of mind, or it may simply be the unauthorized interpolation of the translator.

In addition to his published work, Descartes also collected a significant number of unpublished manuscripts; these were transported to Sweden and, after his death, were sent to Paris. Some of Descartes's papers were subsequently edited and prepared for publication by Claude Clerselier, his literary executor, in the 1660s. They included copies of his letters, which throw much light on his daily work and on his reactions to other books or rival hypotheses; they also confirm the extent to which he was a full-time scientific researcher or natural philosopher. Even more importantly, the unpublished papers contained the first version by Descartes of a comprehensive mechanical worldview, under the title *The World*. Thus any attempt to interpret Descartes's concept of scientific method must take account, not only of his published works, but also of the views that were either withheld from publication during his life or were shared, in confidence, with trusted correspondents.

Finally, it must be acknowledged that Descartes was a practising scientist who both described his scientific theories and provided a commentary on the method by which this scientific work was allegedly guided. If we want to find out what he thought about scientific method, therefore, we cannot simply read what he wrote about it and assume that his commentary is perfectly accurate. In describing the methods he used or the implications, for theory of knowledge, of the new sciences, Descartes was to some extent a captive of the philosophical language and categories that were available in his own day. What he does in science, and what he says he does, may not coincide perfectly. Thus, in looking at Descartes's account of scientific method, one needs to consult both what he wrote about method and how he practised it, and then try to provide a coherent account which includes as much as possible of his whole work.

THE COMPOSITION OF THE *DISCOURSE ON METHOD*

In the history of ideas from Plato and Aristotle to the seventeenth century, there is a very well-established tradition of writing commentaries on Aristotle's *Topics* and *Posterior Analytics*. In the *Posterior Analytics*, Aristotle had defined scientific knowledge as beliefs expressed in the form of a *demonstration*, i.e. first principles that are known to be certain by intuition, and propositions that are deduced from first principles by a series of valid syllogisms. This ideal of what counts as scientific knowledge cast a long shadow of suspicion, for almost twenty centuries, over the value of any discipline that was not structured in this way. It fostered a sharp dichotomy between genuine scientific knowledge, which is certain and is expressed in the form of demonstrations, and mere opinion, which unavoidably fails to achieve the status of reliable knowledge.

Following Aristotle's lead, writing about method became a distinct genre in the centuries that followed; and while many contributors – such as Galen in the case of medicine, or Ptolemy in astronomy – brought their own scientific experience to bear on discussions of method, many others became famous simply as commentators on Aristotle or even on Aristotle's commentators. Thus, for example, Peter Ramus (1515–72) and Jacopo Zabarella (1533–89) were well-known exponents in the sixteenth century of the 'correct' method or methods to be used in arranging items we wish to know in a particular order, or presenting our knowledge in the form of an Aristotelian demonstration. There was nothing unusual or novel, therefore, for a natural philosopher to compose a treatise on method in the early seventeenth century. Francis Bacon wrote a famous essay on method, the *Novum Organon* (1620), in which he proposed a 'great renewal' of knowledge, based on empirical methods. Descartes likewise conceived, in the *Discourse on Method*, of a fundamental renewal of knowledge, and he took his cue from the certainty of mathematical knowledge. The *Discourse* should therefore be seen as another contribution to an already well-established tradition of writing about method. However, long before he wrote the *Discourse*, Descartes embarked on the project of developing a new mechanical physics, of actually doing the science on which the *Discourse* was a commentary.

Between 1619 and 1628, Descartes had begun work on various problems in optics and geometry, and had written brief notes about the possibility of a universal mathematics or a new general method. These notes were collected under the working title: *Rules for Guiding One's Intelligence in Searching for the Truth*. Towards the end of 1628 he retreated from the busy life of Paris, settled in the Netherlands, and dedicated himself full-time to constructing a new physics. He also wrote a short treatise on metaphysics, and began to draft essays on optics, meteorology and geometry. Many of Descartes's ideas for a new physics were crystallized in a book called *The World, or a Treatise on Light*, on which he worked between about 1630 and 1633. When it was ready for publication, he heard news of Galileo's condemnation by the Roman Inquisition and realized that his own account of planetary motions was subject to the same theological objections. He promptly cancelled plans for publication and hid the manuscript of *The World* in his study, where it remained until his death. It was published posthumously, in two parts, in 1662 and 1664.

The relevance of this posthumously published book for understanding the *Discourse* is that, whatever Descartes might have meant by a new scientific method, it had already been applied to basic issues in physical theory in *The World*, and it is unlikely that he had changed his mind completely about how to acquire scientific knowledge in the brief interval between withdrawing *The World* from publication in 1633 and beginning, in 1635, to draft the *Discourse*. *The World* begins with a discussion of the unreliability of our sensations as reflections of the way the world is. The reason given is that there may be a significant difference between our sensation of light and the physical reality which causes that sensation. 'The first thing I want to bring to your attention is that there may be a difference between our sensation of light . . . and whatever it is in the objects that produces that sensation in us.' (p. 85.) Descartes argues that words have meaning, as conventional signs, without resembling what they denote; likewise, the tickling sensation we experience when a feather touches our skin does not resemble any corresponding phenomenon in the feather. Once that is granted, it follows that the natural philosopher cannot deduce the objective nature of various phenomena from our sensations or perceptions; rather, the most we can hope to do is to formulate hypotheses which match our experience of the phenomena in question. Descartes also draws the conclusion that it may be possible to have more than one model of what light actually is, and that a mathematical analysis of its action provides an *explanation* of light, without deciding ultimately between alternative models. Thus, to explain a natural phenomenon is to construct a hypothesis or model, and the success of a hypothesis does not guarantee its uniqueness. Alternative hypotheses may be equally explanatory.

A second feature of the type of scientific explanation adopted in *The World* is that Descartes rejects the forms and qualities that had traditionally been accepted as explanations by scholastic philosophers; or, if he does not reject scholastic explanations completely, he at least claims that he can manage very well without them. For example, in explaining how a piece of wood burns in a flame, Descartes suggests that 'someone else may imagine, if they wish, the "form" of fire, the "quality" of heat, and the "action" which burns it, as things that are completely distinct in the wood. For my part ... I am satisfied to conceive in it the movement of its parts.' (p. 87.) The project of *The World*, then, was to construct in outline a mechanical explanation of light and other related natural phenomena and, in doing so, to provide a new model of what a scientific explanation should look like.

This initiative in physical theory presupposed, for Descartes, a fundamental metaphysical distinction between the physical world and the spiritual world. The human soul, angels and God were classified as spiritual or non-material substances; apart from them, all other realities were classified as physical, and the kind of explanation that is appropriate for physical phenomena is a scientific one. This metaphysical distinction between two radically different types of reality was a question that Descartes had begun to work on soon after his arrival in the Netherlands, and it was summarized in another unpublished draft essay at that time. An expanded version of his metaphysics appeared later in the *Meditations* (1641).

Thus in the years between 1633 and 1636, during which Descartes reached his fortieth birthday, he had accumulated a backlog of unpublished manuscripts in his study. There were preliminary versions

INTRODUCTION

of a general method, an initial outline of a treatise on metaphyics, work on optics and geometry, and two treatises that had been planned as parts of *The World* – a treatise on light, and a treatise on the human body. But, despite all this work and the apparent expectations of his friends that he was ready to reveal the results of his research, he had not yet published anything. It was in these circumstances that, as he explains in Part Six of the *Discourse*, he changed his earlier decision of 1633 and decided to publish, very selectively, some samples of the work that he had been pursuing during the previous eighteen years. One reason for doing so was the requests from friends and admirers. The other reason was the realization that, in order to complete the major renewal of knowledge on which he had embarked, he would need many more observations and experiments than he could hope to complete within the limits of his financial resources or life expectancy. Thus Descartes turned to his voluminous collection of work in progress, and began to rewrite various essays, taking care to exclude the theoretical or contentious assumptions on which they were based. This provided enough material for three essays, the *Dioptrics, Meteors* and *Geometry*. The only thing missing, at this stage, was a preface that might link the essays together and explain how they fitted into a more fundamental project of renewing the sciences. It was for this reason that he began writing the *Discourse on Method*.

Descartes had hoped that the three scientific essays would be published by Elsevier. When this proved unsuccessful, he found another publisher in Leiden, Jan Maire. In March 1636, he described the projected book as '*The Project of a Universal Science which could elevate our nature to its highest level of perfection, etc.*' In it, he planned to 'unveil a part of my method . . .' (p. 57). The publisher's contract was signed in December 1636. Two months later, he still had not finalized the title of the book, but decided to call the introduction a *discourse*, rather than a *treatise* on method, because 'I did not plan to explain my whole method, but merely to say something about it' (p. 58). The following month the same point is made in a letter to Marin Mersenne, a priest who lived in Paris and Descartes's most regular correspondent: 'I do not have *A Treatise on the Method* but *A Discourse on the Method*, which is the same as *A Preface or Notice about the Method*, to show that I do not plan to teach the method here but merely to speak about it.' (p. 59.) So whatever else this introduction was meant to be, it was not intended by its author as a comprehensive and detailed exposition of scientific method.

It is also evident, even on first reading, that the contents of the *Discourse* include items from a rather disparate range of disciplines. There are questions about God and the soul, morality, the circulation of the blood and scientific method, all pieced together in a relatively short essay. To add to the apparent lack of structure, this preface is written continuously from start to finish, although the reader is advised that, if it proves too long to read at one sitting, it may be distinguished into six parts. Descartes's correspondence suggests that he initially considered writing a brief preface about the three scientific essays, and internal evidence in the text suggests that this survives in Part Six of the *Discourse*. A number of correspondents also wrote to Descartes during this period, urging him to publish his physics, i.e. the *World*, but he was still reluctant to do so for obvious reasons. As a compromise with their requests, he seems to have agreed to give some indication of the contents of the unpublished *World*, to whet readers' appetite for more. This explains the otherwise unusual appearance of a lengthy discussion of blood circulation in the preface of a book which contains no physiology. As the preface to the essays began to take shape, he consulted some of his other unpublished essays; these included the *Rules*, an earlier draft of an autobiography, and the first draft of the metaphysical meditations. Thus he tells us in a letter to Father Vatier that the discussion of God and the soul in Part Four was inserted 'only at the last minute, when the publisher was rushing me' (p. 65).

The resulting preface was a reworking of extracts from unpublished works in progress, to indicate the scope of the original project, to provide samples of the variety of topics which would be included (from metaphysics to optics and mathematics), and to encourage readers to believe that the method used by the author was likely, in due course, to be extremely fruitful in resolving questions that had remained unresolved for a very long time. However, the last-minute editing was inadequate to conceal the variety of sources from which different parts of the *Discourse* were borrowed. For, despite the implications of the title, it is not primarily about method; and the brief rules of method included bear little resemblance to the methods illustrated in the accompanying text. The reason for this was that the method proposed both in the *Rules* and in the *Discourse* was a method for discovering new truths, whereas the method appropriate for the exposition of results was rather different. Thus he wrote to Father Vatier, in 1638: 'I was not able to illustrate the use of this method in the three treatises that I have provided, because it prescribes an order for discovering things that is rather different from what I thought should be used in order to explain them.' (p. 65.)

None of these suggestions about the composition of the *Discourse*, however, changes the most obvious feature of the text, namely, that it was written as a preface for the 1637 essays.* Given the context and complexity of its composition, it seems beyond dispute that this brief outline of Cartesian method should be read in the context of the scientific work in support of which it was originally written.

CARTESIAN SCIENTIFIC METHOD

One of the first issues raised by sympathetic readers of the 1637 essays was the status of the initial assumptions or hypotheses that Descartes had used in the *Dioptrics* and the *Meteors*. The famous sceptical arguments of the *Meditations* had not yet appeared in print, nor had the ideal of clear and distinct knowledge with which Descartes promised to provide a definitive refutation of the general scepticism which was widespread in the early seventeenth century. However, he had provided a sufficiently detailed summary of these issues, in Part Four of the *Discourse*, to raise expectations about the level of certainty required for genuine scientific knowledge. But even if the question of certainty had not been raised explicitly, the Aristotelian concept of demonstration was so prevalent in the philosophy of the schools, and was so familiar to readers, that Descartes must have anticipated the question: do these essays provide

---

* Since then it has often been detached from its original context – so much so that some modern commentaries refer to the scientific essays as appendices to the *Discourse*, as if the main contents of a book could be appendices to its preface.

scientific knowledge in Aristotle's sense? Galileo had notoriously come to grief only three years earlier, not so much by proposing a model of the planetary system that differed from what was generally accepted, but by defending the Copernican system as if it were 'demonstrated' knowledge. The response of Cardinal Bellarmine and the Inquisition to Galileo was, in part: there is no objection to the Copernican theory as an astronomical hypothesis or a mathematical model, but it cannot be defended as genuine demonstrated knowledge. The same question was now being addressed to Descartes. Do the scientific essays claim to offer demonstrations, or are they merely hypotheses or models, similar to those of Copernicus, which are no more than merely probable?

Descartes adverts to this question in Part Six, when he acknowledges the ambiguous status of his assumptions. 'If some of the issues that I have spoken about at the beginning of the *Dioptrics* and *Meteors* shock people initially, because I call them assumptions and seem not to want to prove them, they should have the patience to read the whole text attentively . . .' (p. 53.) He goes on to say that, if readers follow this suggestion, they will discover that his description of hypothetical causes is confirmed by their effects, and that the effects are explained by their hypothetical causes. This resolution evidently did not satisfy some critics. Jean-Baptiste Morin, who was a professor of mathematics and astronomy in Paris and a defender of the immobility of the earth, argued that this was circular reasoning, and that it is always possible to imagine some cause to explain any effect one wishes. Descartes's initial response to this kind of objection, in the *Discourse*, was that he could have provided the proof requested, but did not wish to do so for fear of exposing the principles of his physics to the kind of public dispute he was trying to avoid. 'I only called them "assumptions" to let it be known that I think I can deduce them from those first truths that I have explained above.' (p. 53.) This is a rather different answer from what he wrote to Mersenne and Morin the following year, 1638. To Mersenne he wrote: 'to demand geometrical demonstrations from me, in something which depends on physics, is to expect me to do the impossible' (p. 73). And to Morin: 'there is a big difference between proving and explaining. I add that one can use the term "demonstrate" to mean one or the other, at least if it is used according to common usage and not with the special meaning that the philosophers give it.' (p. 75.) None of these replies provided the kind of clarity that Descartes might have implied in the confident and almost dismissive style of his replies. It still appeared to critics as if he promised more than he could deliver – that he claimed to provide demonstrations (but not in the Aristotelian sense) and that he realized, despite his rhetoric, that it was impossible to provide Aristotelian demonstrations in physics. And so he argued that 'for those who simply say that they do not believe what I have written because I deduce it from certain assumptions that I have not proved, they do not know what they are asking for, nor what they ought to ask for' (p. 74).

This was a complex issue in theory of knowledge to which there was no satisfactory answer available in 1637. Descartes was being asked whether he could provide demonstrations which are certain, or whether his hypotheses were merely probable and therefore not genuinely scientific. He seems to have recognized that he could not provide demonstrations in the Aristotelian sense of that term; but he was naturally reluctant to concede that his work was mere guesswork. What was needed, and what did not become available for perhaps another two centuries, was a clear recognition that the most we can hope to achieve in explaining natural phenomena is hypotheses that are more or less plausible. This paradigm shift in theory of knowledge eventually made it possible to concede, without apology, that one need not even attempt to provide the certainty of Aristotelian demonstrations.

The issue of certainty or otherwise was not the only one that emerged from Descartes's reflections in the *Discourse*. It was equally clear that a new concept of explanation was being proposed, one that was ably defended in subsequent decades by Boyle, Huygens and others. As already indicated above, Descartes had suggested in chapter two of *The World* that he did not need the forms and qualities of scholastic philosophy to explain the phenomenon of light, and that the explanatory success of his basic hypotheses far exceeded that of alternative accounts:

If one then compares other people's hypotheses with mine, that is, all their real qualities, substantial forms, elements and things like that, which are almost infinite in number, with the single hypothesis that all bodies are composed of some parts, which is something that can be seen with the naked eye in many cases and can be proved by innumerable reasons in the case of others, ... and if one compares what I have deduced from my assumptions concerning vision, salt, the winds, the clouds, snow, thunder, the rainbow, and similar things, with what others have deduced from their assumptions about the same phenomena, I hope that will be enough to convince those who are not too prejudiced that the effects that I explain have no other causes apart from those from which I deduced them, ... [pp. 76–7]

Here Descartes compares his hypotheses favourably with the lack of success of others, not simply because his own hypotheses were more fruitful, or because he could explain a wide range of apparently disparate phenomena by means of the same hypothesis. He also argues that the forms and qualities of the scholastic tradition are, in a fundamental way, non-explanatory: 'these qualities seem to me to be in need of explanation themselves' (p. 98). In the example used by Boyle, to say of a key that it has 'a lock-opening form' does not explain how or why it opens a lock; it merely renames what needs to be explained in an apparently technical language which is actually no more than an empty and uninformative jargon.

The combined effect of these two arguments, about the superiority of mechanical explanation and the necessity to construct hypothetical models of the causal mechanisms by which natural phenomena occur, showed the extent to which Cartesian science cannot realize the ideal of a demonstrative science proposed by Aristotle. At the same time, it was clear to critics that the results of this new scientific method were not as successful as one might have anticipated from the glowing commentaries of its proponents. In many ways, Descartes and his contemporaries lived through a transitional period – which Thomas Kuhn describes as a paradigm shift – when the lack of success of traditional natural philosophy was evident, but the benefits of the new sciences were not yet clearly available.* There is therefore an element of rhetorical promotion in the accounts of scientific method one finds

---

* Thomas S. Kuhn, *The Structure of Scientific Revolutions*, 3rd edition, University of Chicago Press, 1997.

in the early 1600s. While anxious not to overplay their hand, proponents of the new sciences tried to defend the efficacy of a new, experimental method and, at the same time, to avoid the kind of bruising conflicts with theologians and Christian churches which could result in public censure, imprisonment or worse.

It is not surprising, then, if the *Discourse* appears to the modern reader to be claiming a number of different things which are not obviously compatible: that the author has a new method that can be applied to resolve problems in disciplines as diverse as mathematics and metaphysics; that this method is superior to what was formerly available; that the scientific essays which accompany the *Discourse* illustrate how successful this method is; that the essays rely on initial assumptions about the nature of the phenomena to be explained, and these make it possible to construct mechanical explanations; that such explanations are not Aristotelian demonstrations; that the author could provide proofs of his assumptions, but only by revealing the first principles of his physics, something that he was not yet ready to do; and that those who demand demonstrations in physical science are looking for what is both unnecessary and impossible. This suggests that, if there were important transitions in the history of science between the birth of a new theory and the eventual demise of its predecessor, there were similar transitions in the history of scientific methods. The *Discourse on Method* belongs to one such transition; it marks a decisive turn away from Aristotelian demonstrations towards the hypothetico-deductive strategies of modern science. But the transition was realized with trailing connotations of the methodology that was being replaced. In particular, the continued use of the word 'demonstration' camouflaged the extent to which Cartesian method, when applied to natural phenomena, could not possibly realize what that concept traditionally implied.

THE WORLD AND *THE PRINCIPLES*

The essays of 1637 were presented by Descartes as the first illustrations of his scientific method in operation. One has to look to his later career, therefore, for further applications and for mature expressions of what he meant by scientific knowledge. The *Principles of Philosophy* (1644) is the most comprehensive account of Cartesian physics published during the author's lifetime. Although originally envisaged as a larger enterprise than the published version, the text of 1644 includes four parts. Part One is a revised statement of his metaphysics, which had already been published in greater detail in the *Meditations* (1641). Part Two of the *Principles* is Descartes's general theory of matter, of space and time, together with the three laws of nature and the impact rules for colliding bodies. Parts Three and Four, respectively, provide an extended discussion of Cartesian astronomy and theory of light, and explanations of various terrestrial phenomena such as magnetism. If one asks, in relation to this work, what kind of method is being deployed by Descartes, the most plausible answer is one that reflects the ambiguity already noted in the *Discourse*, viz. that Descartes continues to speak of demonstration, proof and certainty, but that he has accommodated his practice of science to the inevitable uncertainty of hypotheses which are more or less confirmed by supporting evidence.

The most basic assumption of the entire Cartesian project in physics is that all natural phenomena may be explained in terms of the motions and interactions of small parts of matter. Descartes argues that there is only one type of matter in the universe, and therefore that all phenomena that appear in nature must result from combinations of pieces of matter of different sizes, shapes, speeds, etc. In specifying the relevant properties of parts of matter in motion, Descartes was extremely reductionist; for example, there was no room in his physics for electrical or optical properties, or for any attractive forces between particles, that are not reducible to the mechanical interactions of parts of matter. At the same time, he acknowledged that questions about the number of distinct types of particle, their sizes, shapes and motions, cannot be determined by reason on its own. One has to construct the most plausible and most economical hypothesis possible, and see how well it works:

From what has already been said we have established that all the bodies in the universe are composed of one and the same matter, which can be divided into indefinitely many parts ... However, we cannot determine by reason alone how big these pieces of matter are, or how fast they move, or what kinds of circle they describe. Since there are countless different configurations that God might have instituted here, experience alone must teach us which configurations he actually selected in preference to the rest. We are thus free to make any assumption on these matters with the sole proviso that all the consequences of our assumption must agree with our experience. [AT VIIIA, 100–101; CSM I, 256–7]

In looking back at assumptions like this from the perspective of twentieth-century physics, it is not obvious why some natural philosophers of this period were so parsimonious in specifying the basic properties of parts of matter. One suggestion is that they adopted Occam's principle – which implied that, in constructing a theory, one should not postulate any more variables than are absolutely necessary. Accordingly, one should begin by describing particles of matter with as few properties as possible and see how successful the theory is. Given the prodigality of their scholastic critics, it seems plausible to understand the strategy of the new natural philosophers as a reaction to the almost unlimited forms and qualities of traditional philosophy. There was also a second reason for this strategy: they did not understand any properties of matter apart from their size, shape and motion, and therefore other properties that one might have wished to postulate would not have been amenable, at the time, to mathematical modelling.

This motivation is clear in some of Descartes's comments. Many of the particles of matter that were postulated to explain observable phenomena, such as the screw-shaped particles imagined as the causes of magnetism, were unobservable to the naked eye, and they remained unobservable even with the assistance of early microscopes. Thus Cartesian science was unashamedly dealing with unobservable particles. In this situation, Descartes suggests to Plempius – one of his Dutch correspondents, who was concerned about this strategy – that 'there is nothing more in keeping with reason than to judge about those things that we do not perceive, because of their small size, by comparison and contrast with those that we see' (AT I, 421). The same kind of answer is given in reply to objections from Morin, in 1638: 'in the analogies I use, I only compare some movements with others, or some shapes with others, etc.; that is to say, I compare those things which, because of their small size, are not accessible to our senses with those that are perceptible; they do not differ from the latter more than a small circle differs from a large one' (AT II, 367–8).

Having opted for mechanical explanations, and having agreed to postulate whatever small particles of matter seem to be required to explain the range of natural phenomena that are discussed in the *Principles*, Descartes's most basic theoretical need was to formulate the laws that describe the motions and interactions of such particles. This was realized by three laws of nature and seven impact rules. The way in which the laws and rules were justified is a complex question. Descartes typically overstates his case by claiming to have proved or demonstrated them; and he certainly gives the impression, both in *The World* and in the *Principles*, that they are derived exclusively and deductively from metaphysical foundations, such as the axiom that God is the ultimate cause of all effects in the universe, and that his actions are eternal and unchanging. However, one might equally interpret the link between metaphysics and the laws of nature less deductively. On that reading, Descartes merely required that the laws be consistent with his metaphysical principles, and that they be justified, at least partly, by the natural phenomena they explain. The ambivalent claim, that the laws of nature are derived from a metaphysical foundation and that they are so successful that they must be accepted as true, is repeated at the conclusion of the *Principles*:

there are some matters ... which we regard as absolutely certain, and more than just morally certain ... This certainty is based on a metaphysical foundation ... Mathematical demonstrations have this kind of certainty, as does the knowledge that material things exist; and the same applies to all evident reasoning about material things. Perhaps even these results of mine will be allowed into the class of absolute certainties, if people consider how they have been deduced in an unbroken chain from the first and simplest principles of human knowledge ... it seems that all the other phenomena, or at least the general features of the universe and the earth that I have described, can hardly be explained intelligibly except in the way that I have suggested. [AT VIIIA, 328–9]

This suggestion, that relatively few and simple hypotheses have been fruitful in explaining a wide range of apparently disparate natural phenomena, including many for which the hypotheses had not originally been devised, is often invoked as a confirming justification:

> Now if people look at all the many properties relating to magnetism, fire and the fabric of the whole world that I have deduced in this book from just a few principles, then even if they think that my assumption of these principles was arbitrary and groundless, they will still perhaps acknowledge that it would hardly have been possible for so many items to fit into a coherent pattern if the original principles had been false. [AT VIIIA, 328]

Whatever conclusion one reaches about Descartes's claims on behalf of the basic principles of his physics, including the three laws of nature, there are unambiguous indications in the *Principles* that, in some cases, we must be willing to settle for hypotheses that are uncertain and even, occasionally, for hypotheses that we believe are false. The reason for this was that we would be better off with some model of how a phenomenon may have been mechanically caused, than with none at all.

> As far as particular effects are concerned, whenever we lack sufficient experiments to determine their true cause, we should be content to know some causes by which they could have been produced . . . I believe that I have done enough if the causes that I have explained are such that all the effects that they could produce are found to be similar to those we see in the world, without inquiring whether they were in fact produced by those or by some other causes. [AT IXB, 185, 322]

Thus the *Principles* confirms the trend in Descartes's thinking which first appears in the *Discourse*. There are a few basic laws of nature that depend, in some sense, on the metaphysical foundations provided by the *Meditations*, and they enjoy the same degree of certainty as those foundations. Once we go beyond these and begin to construct mechanical explanations of natural phenomena, we cannot avoid constructing hypotheses; for the particles of matter used to explain natural phenomena are often unobservable and, even if we could observe them, there is no guarantee that the objects of our perceptions resemble our sensations of them. These hypotheses in turn are confirmed by the extent to which they explain natural phenomena, and we can be more confident about them when relatively few and simple hypotheses succeed in explaining a wide range of apparently disparate phenomena, such as the daily motion of the tides, the effects of magnetic stones, the circulation of blood, or the refractive effects of lenses on light. While Descartes still claims mathematical certainty for some of his fundamental principles, he is willing at last – even if only reluctantly – to settle for probability or moral certainty in the case of many scientific explanations.

THE *RULES FOR GUIDING ONE'S INTELLIGENCE*

If we adopt this account of scientific method from Descartes's mature writings, there is then a question about the significance of the *Rules* as an early expression of his method. The *Rules* were composed at various times between about 1619 and 1628, and they reflect Descartes's interests at different stages of their composition within that period. In his early work with the Dutch physicist, Isaac Beeckman, in 1618–19, Descartes was primarily interested in the possibility of a universal *mathesis*. Subsequently, he drafted rules for discovering new truths – an early version of a logic of discovery – and he also outlined a general account of our cognitive faculties. All these issues feature at different places in the text. The plan of the book, at the time when Descartes gave up working on it, was to write three sets of twelve rules. But only twenty-one rules survived, the last three of which are incomplete. If we read the *Rules* today and ask how significant they are as a statement of Cartesian method, it is necessary to acknowledge the fact that they were composed when Descartes was merely beginning to formulate a new theory of science. It is also plausible to think that the *Rules* were intentionally abandoned, rather than that they were simply left incomplete, and that Descartes had good reason to abandon them. What those reasons may have been depends very much on one's interpretation of the extant sections of the text.

One possibility is to think of Descartes's work here as comparable to the early logical treatises of Galileo's student days, in which the young Galileo copied into a notebook standard accounts of Aristotelian method in science and appropriated them to his own use with the limited originality of a young student.* Descartes may have similarly conceived of a general method that could be used to solve problems in mathematics, optics, magnetism, etc. The assumption that there might be a single method or set of rules that applies to many different disciplines was already familiar in methodological treatises in the Renaissance. The idea that the application of a method could produce the certainty of intuition or deduction was plausible in the case of mathematical proofs, and was not very different from what Aristotle had specified for scientific knowledge in the *Posterior Analytics*. But it is also plausible to think that, once Descartes began to do scientific experiments, he quickly realized that scientific explanations are much more hypothetical and uncertain than his *Rules* anticipated, and that a different approach was needed if he were to make the kind of revolutionary contribution to the experimental sciences that was announced in the original title of the *Discourse*.

However, even if there was a significant change of plans between 1628 and 1637, there are already some indications in the *Rules* that the method proposed by Descartes was not as unequivocally a priori and deductive as might initially appear. It is true that he contrasts the certainty of intuition and deduction with the deceptiveness of experience, and that he highlights the way in which a merely probable claim, when included in a deduction, compromises the certainty of the conclusion. He also gives a prominent role to 'certain seeds of truth which are innate in the human mind' (p. 128). At the same time, however, the *Rules* propose that 'the intellect alone is capable of perceiving the truth, but it must be assisted by the imagination, sensation and memory so that we do not happen to omit anything that was provided among our powers' (p. 151).

Descartes also seems to have recognized, even at this early stage of his career, that there is a significant difference between purely mathematical questions and questions that belong to physics. Thus in

* See William A. Wallace, *Galileo's Logical Treatises*, Kluwer, Dordrecht, 1992.

his discussion of the anaclastic in Rule Eight, he points out that anyone whose studies were limited to mathematics 'will realize easily that . . . the determination of this line depends on the ratio between the angles of incidence and the angles of refraction. But . . . they will not be able to pursue this further because it belongs to physics rather than to mathesis.' (p. 139.)* The strategy proposed for escaping from the limitations of pure mathematics depended ultimately on understanding what is meant by a 'natural power', and this in turn was to be accomplished by reflecting on our experience of local motions. Failure to consult such everyday experiences is a breach of Rule Five; among those who break the rule are 'philosophers . . . who, neglecting experience, think that the truth will spring from their own brains as Minerva did from the head of Jove' (p. 131).

The role of experience in physics, in contrast with pure mathematics, is reflected in the distinction between two kinds of problem, perfect and imperfect, which were planned for discussion in the projected rules 13–24 and 25–36. 'As regards questions, some may be understood perfectly, . . . and we discuss those alone in the next twelve rules. There are others, finally, that are not perfectly understood, and we defer them to the final twelve rules.' (p. 164.) By 'perfect questions' Descartes understood problems whose solution could be derived logically from the knowledge we already have. That suggests that the type of question involved in imperfect problems is sufficiently different to warrant a separate strategy or a different set of rules. It is true that, in Rule Thirteen, Descartes claims that all imperfect questions may be reduced to perfect questions: 'From these data it is easily understood how all imperfect questions can be reduced to perfect questions.' (p. 165.) But even this apparently reductionist ideal is qualified by the example he gives of what such a reduction looks like in practice. To explain magnetism, the only option available to human intelligence is one that builds on various natures that are already known to us. But,

if there is some kind of thing in the nature of the magnet such that our intellect has never perceived anything similar to it, it cannot be expected that we would

---

* The same distinction is made in Rule Twelve, p. 152.

ever get to know it by reasoning; we would have to be taught it either by some new sensation or by the divine mind. But if we perceive distinctly the combination of already known things or natures which produce the effects that appear in the magnet, we shall consider that we have learned whatever can be provided in this context by human intelligence. [p. 170]

That implies that knowledge of the relevant basic natures that are needed to construct a physical explanation can be acquired only from sensation (unless God intervenes to instruct us directly). It also implies that we will have gone as far as is humanly possible if we invent a hypothesis, based on what we already know from experience, which could account for the observed phenomena. That is what Descartes apparently means by making imperfect questions amenable to the scientific method outlined in the first twelve rules.

If it is true that the *Rules* were abandoned, perhaps because scientific method in physics or physiology turned out to be even more different from mathematics than Descartes had realized, it was still possible to include a remnant of the ideal of a universal method in the four summary rules of the *Discourse* (Part Two). They imply that the author had a new method of discovery, that it was applicable to a wide range of disciplines, from metaphysics to geometry, and that the first fruits of its application were now available to readers for inspection. But the method actually implemented in the scientific essays of 1637, and subsequently in the *Principles*, had little in common with the sketchy proposals of ten years earlier.

TEXTS IN THIS EDITION

The focus of this selection of texts is Descartes's contribution to a new concept of science in the early decades of the seventeenth century. Among the texts available, the *Discourse on Method* remains the most accessible and most famous of his discussions of scientific method. Evidently, it would be preferable to read this in association with the essays for which it was written as a preface; but the *Dioptrics, Meteors and Geometry* are rather lengthy essays in early modern science which are not easily abridged. Luckily, Descartes's unpublished text of *The World* did survive its famous journey to Paris, after the author's death, including a brief visit to the bottom of the river Seine; it includes a programmatic and readily accessible version of his whole physics. I have included the first seven chapters of *The World* here, as a sample of the kind of scientific renewal that Descartes had in mind when writing the *Discourse*, and as an illustration of his scientific method in practice in the 1630s.

I have also included selections from Descartes's correspondence during the years immediately prior to publishing the *Discourse*, and his replies to some critics of his method after 1637. These letters, although they represent a very small selection from the hundreds which survived, provide enough evidence to show that their author did not set out to write a coherent and comprehensive account of scientific method in 1637. It is obvious, instead, that Descartes was oscillating between revealing and concealing some of the unpublished contents of his bureau, and the edited final product shows the effects of haste in its composition under pressure of a publisher's imminent deadline.

Finally, students of Descartes will continue to dispute the significance of the *Rules* for the interpretation of his scientific methodology. It may be read, as already indicated, as an early draft of ideas that were superseded by the experience of doing scientific work in areas such as optics and anatomy. Alternatively, it could be read as a programmatic statement of an ideal to which Descartes remained committed in principle during his subsequent work, although he adjusted it more or less significantly in the light of his scientific experience. For that reason, I have included the full text of the *Rules*, so that readers can make up their own mind on the question.

Further Reading

Armogathe, J. R. and Marion, J. L., *Index des Regulae ad Directionem Ingenii de René Descartes*, Editioni dell' Ateneo, Rome, 1976.

Belgioioso, Giulia, et al., eds., *Descartes: il metodo et i saggi*, Istituto della Enciclopedia Italiana, Rome, 1990.

Bitbol-Hespériès, Annie, *Le Principe de vie chez Descartes*, Vrin, Paris, 1990.

Broughton, Janet, *Descartes's Method of Doubt*, Princeton University Press, Princeton and Oxford, 2002.

Campanella, Thomas, *A Defense of Galileo the Mathematician from Florence*, trans. Richard J. Blackwell, University of Notre Dame Press, Indiana, 1994.

Clarke, Desmond M., *Descartes' Philosophy of Science*, Manchester University Press, Manchester, 1982.

Clarke, Desmond M., *Occult Powers and Hypotheses: Cartesian Natural Philosophy under Louis XIV*, Clarendon Press, Oxford, 1989.

Clarke, Desmond M., *Descartes's Theory of Mind*, Clarendon Press, Oxford, 2003.

Cottingham, John, ed., *The Cambridge Companion to Descartes*, Cambridge University Press, Cambridge and New York, 1992.

Cottingham, John, *A Descartes Dictionary*, Blackwell, Oxford, 1993.

Crapulli, Giovanni, ed., *Descartes: Regulae ad directionem ingenii*, Nijhoff, The Hague, 1966.

Descartes, René, *Meditations and Other Metaphysical Writings*, trans. D. M. Clarke, Penguin Books, Harmondsworth, 1998.

Descartes, René, *The Philosophical Writings of Descartes*, trans. J. Cottingham, R. Stoothoff, D. Murdoch and A. Kenny, 3 vols., Cambridge University Press, Cambridge, 1985–91.

Garber, Daniel, *Descartes' Metaphysical Physics*, University of Chicago Press, Chicago and London, 1992.

Garber, Daniel, *Descartes Embodied*, Cambridge University Press, Cambridge, 2001.

Gaukroger, Stephen, ed., *Descartes: Philosophy, Mathematics and Physics*, Harvester Press, Sussex, 1980.

FURTHER READING

Gaukroger, Stephen, *Cartesian Logic: An Essay on Descartes's Conception of Inference*, Clarendon Press, Oxford, 1989.

Gaukroger, Stephen, *Descartes: An Intellectual Biography*, Clarendon Press, Oxford, 1995.

Gaukroger, Stephen, ed. and trans., *Descartes: The World and Other Writings*, Cambridge University Press, Cambridge, 1998.

Gaukroger, Stephen, *Descartes' System of Natural Philosophy*, Cambridge University Press, Cambridge, 2002.

Gaukroger, Stephen, et al., eds., *Descartes' Natural Philosophy*, Routledge, London and New York, 2000.

Gilbert, Neal Ward, *Renaissance Concepts of Method*, Columbia University Press, New York and London, 1960.

Grosholz, Emily R., *Cartesian Method and the Problem of Reduction*, Clarendon Press, Oxford, 1991.

Marion, Jean-Luc, *Règles utiles et claires pour la direction de l'esprit en la recherche de la vérité*, Nijhoff, The Hague, 1977.

Rodis-Lewis, Geneviève, *Descartes: His Life and Thought*, trans. Jane M. Todd, Cornell University Press, Ithaca and London, 1998.

Shea, William R., *The Magic of Numbers and Motion*, Science History Publications, Canton, Mass., 1991.

*Discourse on the Method
for Guiding One's Reason and Searching for
Truth in the Sciences*

1637

NOTE ON THE TEXT AND TRANSLATION

Descartes's first book was published by Jan Maire at Leiden, in 1637. The title-page of the first edition, which did not include the author's name, was: *Discours de la methode pour bien conduire sa raison, & chercher la verité dans les sciences. Plus la Dioptrique, les Meteores, et la Geometrie, qui sont des essais de cete methode.* The *Discourse* was paginated separately (pp. 3–78), as a preface or introduction to the main body of the text, which was paginated from pp. 1–413. This translation is based on the first edition. In preparing the translation I have consulted the Latin version of 1644, *Dissertatio de Methodo recte regendae rationis, & veritatis in scientiis investigandae* (Amsterdam, 1644), and the first English translation, *A Discourse of a Method for the well-guiding of reason, and the Discovery of Truth in the Sciences* (London, 1649).

As indicated by Descartes in the opening paragraph, the text was printed without chapters or any comparable divisions into sections. However, the reader was invited to divide it into parts if it proved too long to be read in a single sitting, and there were indications on the margin of the text where each new part might begin. In deference to this compromise on the author's part, I have included the traditonal subtitles of 'Part One', 'Part Two', etc., but have put them in parentheses to remind readers of their original marginal status.

The text seems to have been composed by Descartes as the printer was completing work on the three essays. For this reason I did not attach any philosophical significance to the fact that some of Descartes's paragraphs are extremely long by modern standards; instead, I provided a few extra paragraph indentions to make the text more readable. Finally, in translating the title, the word '*sciences*' in French causes particular problems. Today we draw a clear distinction between the experimental sciences and the humanities, and the word 'science' in

English has a definite meaning in ordinary usage. However, Descartes is still using the word in a much wider sense to mean something like 'reliable knowledge based on first principles'. In this sense, theology and philosophy were classified as sciences in the early Modern period.

DISCOURSE ON METHOD

If this discourse seems too long to be read all together at one time, it may be divided into six parts. In the first part there are various considerations about the sciences; in the second, the principal rules of the method which the author had looked for; in the third, some of the rules of morality that he derived from this method; in the fourth, the arguments by which he proves the existence of God and of the human soul, which are the foundations of metaphysics; in the fifth, the order of the questions in physics that he has researched and, especially, the explanation of the heart's motion and of some other problems that belong to medicine, and also the difference between our soul and that of brute animals; in the final part, the things that he thinks are necessary in order to make more progress than he has made in the investigation of nature, and the reasons why he wrote this discourse.

[PART ONE]

Common sense[1] is the best distributed thing in the whole world. Everyone thinks that they are well endowed with it, so that even those who are most difficult to please in every other respect do not usually wish to have more than they already possess. It is unlikely that everyone is wrong about this. It shows, rather, that the ability to judge well and to distinguish what is true from what is false – which, strictly speaking, is what is meant by 'common sense' or 'reason' – is naturally equal in all human beings. Thus the diversity of our views does not result from the fact that some people are more reasonable than others, but simply from the fact that we guide our thoughts along different paths and do not think about the same things. For it is not enough to have a good mind; it is more important to use it well. The greatest souls are just as capable of the greatest vices as of the greatest virtues, and those who move only very slowly may make much greater progress if they always follow the right path than those who run but stray from it.

In my own case, I have never presumed that my mind was in any way more perfect than that of the average person. Indeed I have often wished that my thoughts were as quick, or my imagination as clear and distinct, or my memory as ample or prompt as some others. I do not know of any other qualities apart from these that serve to perfect the mind. For in relation to reason or sense, in so far as it is the only thing that makes us human and distinguishes us from brute animals, I believe that it is present in each of us in its entirety, and thus I follow the view generally held by philosophers, who say that there are differences of degree only between 'accidents', but not between the 'forms' or natures of individuals in the same species.[2]

But I have no hesitation in saying that I think I have been very fortunate, ever since my youth, in finding myself on certain paths that led me to thoughts and principles from which I constructed a method by which, it seems to me, I have a way of increasing my knowledge by degrees and raising it gradually to the highest level that the mediocrity of my mind and the brevity of my life allow. I have already derived such benefits from it that, even though I always try to lean towards caution rather than presumption in judgements I make about myself and even if, when casting a philosopher's eye over the variety of actions and undertakings of humankind, there seems to be none among them which is not vain and useless, I still get the greatest satisfaction from the progress that I think I have already made in the search for truth, and I have such hopes for the future that if, among purely human occupations, there is one that is genuinely good and worth while, I am bold enough to believe that it is the one that I have chosen.

However, I may be mistaken, and what I think is gold and diamonds may be merely pieces of copper and glass. I realize how much we may be mistaken about things that are important to us, and how much the judgements of our friends should be distrusted when they are favourable. But I shall be very anxious to indicate, in this discourse, the paths that I have followed, and to represent my life as if it were a painting, so that everyone can judge it for themselves; if I learn what people think of it from the public's response, this will be another way of instructing myself which I shall add to those that I normally use.

Thus my plan here is not to teach the method that everyone must follow in order to guide their reason, but merely to explain how I have tried to guide my own. Those who set themselves up to instruct others must think they are better than those whom they instruct, and if they misguide them in the slightest they can be held responsible. But, since I am proposing this work merely as a history or, if you prefer, a fable – in which, among a number of examples that may be imitated, there may also be many others where it would be reasonable not to follow them – I hope it will be useful for some readers without being harmful to others, and that everyone will be grateful for my frankness.

I have been nourished by books since I was a child, and because I was convinced that, by using them, one could acquire a clear and certain knowledge of everything that is useful for life, I had a great desire to study them. But as soon as I had concluded the course of studies at the end of which one is usually admitted to the ranks of the learned, I changed my mind completely. For I found myself so overcome by so many doubts and errors that I seemed to have gained nothing from studying, apart from becoming more conscious of my ignorance. Despite that, I was at one of the most renowned schools in Europe,[3] where I thought there ought to have been some learned men if there were some anywhere on earth. While I was there, I learned everything that the other students were learning, and since I was not satisfied with the sciences that we were taught, I even looked through all the books I could find that were concerned with what are thought to be the most insightful and unusual sciences. At the same time, I was aware of other people's opinion of me, and I did not notice that they thought I was inferior to my fellow students, who even included some who were already destined to replace our teachers. Finally, the period in which we lived seemed just as developed, and as productive of good minds, as any previous age. That made me feel free to think for myself about all earlier doctrines, and to believe that there was no doctrine in the world of the kind that I had been led to expect.

However I still valued the exercises that occupy people in the schools. I knew that the languages learned there are necessary in order to understand classical texts; that the politeness of fables animates the mind; that the memorable deeds of history uplift it and, when read critically, that they help to train our judgement; that the reading of all good books is like a conversation with the most eminent people of past centuries, who were their authors, and that it is even a studied conversation in which they reveal to us only the best of their thoughts; that oratory has incomparable powers and attractions; that poetry has very ravishing delights and sensibilities; that mathematics contains very subtle discoveries that can help very much to satisfy those who are curious, to facilitate all crafts, and to reduce human labour; that moral writings contain many instructions and many encouragements to virtue which are very useful; that theology teaches us how to get to heaven; that philosophy provides ways of speaking plausibly about everything, and of making oneself admired by those who are less educated; that law, medicine and the other sciences bring honour and riches to those who practise them; and finally, that it is good to have studied them all, even the most superstitious and false among them, in order to know their real value and to protect oneself against being deceived by them.

But I thought I had already devoted enough time to languages and even to reading the classics, to their stories and fables, because conversation with people from other periods is like travelling. It is helpful to know something about the customs of different peoples in order to make a more sensible judgement about our own, and not to think that everything that is different from our ways is ridiculous and irrational, as is usually thought by those who have seen nothing else. But if one spends too much time travelling, one eventually becomes a stranger in one's own country; and if one is too curious about things that happened in past ages, one usually remains very ignorant about what is currently taking place. Moreover, fables make us think that many things are possible when they are not, and even the most accurate histories, although they do not change or exaggerate the significance of things in order to make them more worth reading about, at least almost always omit the less important or significant details; thus what remains does not appear as it really is, and those who regulate their lives by examples drawn from history are in danger of falling into the

[PART ONE]

extravagancies of the knightly heroes of romantic novels, and of thinking up projects that are beyond their abilities.

I had great respect for oratory and I loved poetry, but I thought that both of these were mental gifts rather than the results of study. Those who have the most powerful reason, and who best arrange their thoughts so as to make them clear and intelligible, can always convince others better of what they propose even if they speak only the Breton language and have never learned rhetoric;[4] and those who have the most pleasant compositions and who know how to express them most attractively and ornately, would still be the best poets even if they knew nothing about poetic theory.

Above all else, I was interested in mathematics because of the certainty and self-evidence of the way it reasons; but I had not yet noticed its real use and, since I thought it was useful only for mathematical applications, I was surprised that nothing more noteworthy had been built on such solid and firm foundations. In contrast, I compared the writings on morality of the ancient pagans with very magnificent and superb palaces that were built only on sand and mud. They very much extol the virtues and make them appear more valuable than anything else in the world; but they do not provide adequate instruction about how to recognize them, and frequently something that they classify as a virtue is merely insensibility, pride, despair or parricide.

I respected our theology and hoped, as much as anyone else, to get to heaven.[5] But once I learned, as something which is very certain, that the path to heaven is just as open to the most ignorant as to the most learned, and that the revealed truths which lead there are beyond our understanding, I did not dare subject them to the feebleness of my reasoning, and thought that one needed to have some extraordinary assistance from heaven and to be more than human in order to study them successfully.

I shall say nothing about philosophy, except that it has been practised by the best minds that have appeared over many centuries, and yet it still contains nothing that is not disputed and consequently doubtful; therefore I was not so presumptuous as to hope to succeed better in it than others. And when I considered how many different opinions there may be about the same thing which are defended by the learned, even though no more than one of them can ever be true, I regarded almost as false everything that was merely probable.

Thus, as regards the other sciences, in so far as they borrow their principles from philosophy, I judged that it was impossible that anything solid could have been built on foundations that were so weak, and neither the honour nor the profit that they promised were enough to persuade me to study them. For, thank God, my situation was not such that I had to earn a living from the sciences in order to supplement my income. And although I did not claim to despise fame, like a Cynic, I nevertheless had little respect for what I could acquire only on false pretences. Finally, as far as false doctrines are concerned, I thought that I already knew their value well enough not to be any longer subject to being deceived by the promises of an alchemist, the predictions of an astrologer, the deceptions of a magician, or the tricks and boasts of any of those who claim to know more than they really do.

That is why, as soon as I was old enough to leave the control of my teachers, I gave up completely the study of the humanities and, resolving not to search for any other science apart from what could be found in myself or in the great book of the world, I spent the remainder of my youth travelling, visiting courts and armies, meeting people of different temperaments and rank, acquiring different experiences, testing myself in meetings that came my way by chance, and everywhere reflecting on the things I observed so as to derive some benefit from them. For it seemed to me that I could find much more truth in the reasoning that each person does about things which are important to them, and which have harmful consequences for them if they misjudge, than in those made by a scholar in their study about speculative matters that have no consequences and whose only effect on them, perhaps, is that the further removed they are from common sense the more vain they will be about them, because they would have to use so much more ingenuity and skill in trying to make them plausible. I always had a great desire, also, to learn to distinguish what is true from what is false, in order to see my way clearly in actions and conduct myself with confidence in this life.

It is true that, as long as I was thinking only about the customs of other people, I found hardly anything convincing, and I noticed among them as much diversity as I had previously noticed among the views of philosophers. Thus the greatest benefit I got from this was that, by seeing many things which, even though they seemed very extravagant and ridiculous to us, were still widely accepted and approved by other great peoples, I learned not to believe anything too firmly about which I had been convinced by example and custom alone. Thus I was gradually freed from many errors that can cloud our natural light and make us less capable of hearing reason. But once I had spent some years studying in this way in the great book of the world and trying to acquire some experience, I decided one day to study also within myself, and to use all the powers of my mind to choose the paths that I should follow. I was much more successful in this, it seems to me, than I would have been had I never left either my country or my books.

[PART TWO]

I was then in Germany, where I had been drafted because of the wars that are still going on there,[6] and as I was returning to the army from the emperor's coronation, the arrival of winter delayed me in quarters where, finding no company to distract me and, luckily, having no cares or passions to trouble me, I used to spend the whole day alone in a room that was heated by a stove, where I had plenty of time to concentrate on my own thoughts.[7] Among these thoughts, one of the first that I examined was that there is often less perfection in works composed of several parts, and made by the hands of a variety of contributors, than in those on which only one person has worked. Thus one notices that buildings that were started and completed by a single architect are usually more attractive and better designed than those which a number of architects have tried to put together by making use of old walls that had been built for different purposes. Likewise ancient cities, which were originally only small villages and developed over time into large towns, are usually so badly laid out in contrast with symmetrical town squares designed without restriction by an engineer for an empty site that, even though their buildings, when considered separately, are often as artistic or even more artistic than others, nevertheless when one considers how they are arranged – with a large building here, a small one there – and how they make the streets crooked and irregular, one would say that it was chance rather than a human will using reason that laid them out in this way. However, if one also considers that there have always been officials who were responsible for making sure that private buildings contribute to the embellishment of public places, one will see that it is difficult to achieve very good results merely by adapting what others have produced. In the same way I imagined that peoples who had formerly been half savage and became civilized only gradually, and who had made their laws only in so far as they were forced to do so as a result of problems caused by crimes and quarrels, could never be as well governed as those who, from the time when they first assembled as a community, had observed the basic laws of some wise legislator, just as it is very certain that the state of the true religion must be incomparably better regulated than that of all others, since God alone made its rules. And speaking of human affairs, I believe that if Sparta was formerly very flourishing, that was not because each of its laws taken separately was good – for some of them were very strange and even contrary to good morals – but because they were all devised by a single person and therefore all tended towards the same goal. Thus I thought that the sciences found in books – at least those which are only probable and do not contain any demonstrations, since they were composed and developed gradually from the views of many different people – do not come as close to the truth as the simple reasoning that a person with common sense can perform naturally about things that they observe. I also thought that, since we were all infants before we became adults, and since we were necessarily governed for a long time by our appetites and our teachers, which were often at odds with each other and of which, perhaps, neither always gave us the best advice, it is almost impossible for our judgements to be as clear and as well-founded as they might have been had we had the full use of our reason from the day we were born and had we never been guided by anything else.

It is true that we do not see people knocking down all the houses in a town for the sole purpose of rebuilding them in a different way and making the streets of the town more attractive; but we certainly do see many people demolishing their own houses in order to rebuild them, and they are even forced to do so, in some cases, when their foundations are not very secure and they are in danger of falling down of their own accord. This example convinced me that it would not be reasonable for someone who plans to reform a state to change everything from the foundations up, and to demolish it in order to rebuild; nor would it be reasonable to reform the body of the sciences or the curriculum established in the schools for teaching them. But, in the case of all the views that I had previously accepted among my beliefs, I could do no better than to undertake to remove them all at once, so as subsequently either to replace them with better ones, or with the same ones if they were tested against the criterion of reason. I firmly believed that, in this way, I would succeed in guiding my life much better than if I had built only on old foundations, and had relied only on principles that I allowed myself to be convinced of in my youth without ever checking to see if they were true. For, although I noticed various difficulties in this, they were not irremediable, nor were they comparable to those that occur in the smallest adjustments that affect public life. Large bodies, once they are knocked down, are too difficult to rebuild and even, once shaken, are too difficult to support, and their fall is inevitably very hard. Besides, if they have any imperfections – their diversity alone assures us that they have many – usage has undoubtedly smoothed them over very much, and has even imperceptibly prevented or corrected a number of them for which one could not have provided by being prudent. Finally, it is nearly always better to tolerate them rather than to change them, in the same way that the principal paths which meander between mountains become gradually so smooth and convenient from frequent use, that it is much better to follow them than to try to travel more directly by climbing over rocks and descending to the bottom of precipices.

That is why I could never approve of those confused and anxious people who were destined neither by birth nor fortune for public administration but are still always thinking up some novel reform. If I thought there was the slightest thing in this book by which anyone could suspect me of this folly, I would be very reluctant to allow its publication. My objective was never anything more than an attempt to reform my own thoughts and to build on a foundation that was entirely my own. If I am satisfied enough with my work to show you a sketch of it here, it is not the case that, in doing so, I advise anyone else to imitate it. Those who have a better share of God's gifts will perhaps have higher ambitions; but I fear that even this one may already be too demanding for many people. Even the resolution to give up all the views that one has previously believed is not an example that everyone should follow. There are hardly more than two kinds of minds in the world for whom it is completely inappropriate. There are those who think they are more competent than they really are and thus cannot avoid rushing into judgements, and do not have enough patience to guide all their thoughts in an orderly fashion; the result is that, if they ever took the liberty to doubt the principles that they accepted and to stray from the usual path, they would never be able to stick to the path that must be followed in order to move forward more directly, and they would remain lost all their lives. Secondly, there are those who are sufficiently reasonable and modest to realize that they are less competent to distinguish between what is true and what is false than others who could instruct them; they should be much more content to follow the views of these others than to look for better ones themselves.

As far as I am concerned, I would surely have been among the second group just mentioned if I had never had more than one teacher, or if I had not known about the differences of opinion that have always obtained among the most learned. But from the time I was in college I learned that there is nothing one could imagine which is so strange and incredible that it was not said by some philosopher; and since that time, I have recognized through my travels that all those whose views are different from our own are not necessarily, for that reason, barbarians or savages, but that many of them use their reason either as much as or even more than we do. I also considered how the same person, with the same mind, who was brought up from infancy either among the French or the Germans, becomes different from what they would have been if they had always lived among the Chinese or among cannibals, and how, even in our clothes fashions, the very thing that we liked ten years ago, and that we may like again within the next ten years, appears extravagant and ridiculous to us today. Thus our convictions result from custom and example very much more than from any knowledge that is certain. However, a majority of votes has no validity as a proof for truths that are a little difficult to discover, because it is much more likely that such truths will be discovered by an individual rather than a whole people. Thus, I was unable to choose someone whose views seemed to me to be preferable to those of others, and I found myself forced to take on the task of guiding myself.

But, like someone who walks alone in the dark, I decided to go slowly and to be so careful about everything that, even if I made very little progress, I would at least prevent myself from falling. I did not even wish to begin rejecting completely any of the views that may have slipped among my beliefs previously without having been introduced there by reason, until I had first taken enough time to plan the project that I was undertaking and to search for the correct method for acquiring knowledge of all the things that my mind would be capable of knowing.

When I was younger, I had studied a little logic as part of philosophy and, in mathematics, I had studied geometrical analysis and algebra – three arts or sciences that seemed as if they ought to contribute something to my project. But when I studied them I noticed that, in the case of logic, its syllogisms and most of its other rules are more useful for explaining to someone else what one already knows or even, in the Lullian art, for speaking uncritically about things that one does not know, than for learning them.[8] But even if logic includes many rules that are very true and very good, there are so many others mixed in with them which are either harmful or superfluous that it is almost as difficult to separate them as to extract a Diana or Minerva from a block of marble that is not even roughly hewn. As regards the analysis of the ancients or the algebra of the moderns, apart from the fact that they apply only to very abstract questions which seem to have no use, the former is always so tied to the discussion of shapes that it cannot exercise the understanding without greatly tiring the imagination; and in the latter, one is so constrained by certain rules and symbols that it has become a confused and obscure art that hinders the mind, rather than a science that assists it. That is why I thought I ought to look for another method which would include the benefits of these three, but would be free from their defects. And since a proliferation of laws often provides an excuse for vice, because a state is governed much better when it has only very few laws that are observed very strictly, I believed that, instead of the multiplicity of rules that comprise logic, I would have enough in the following four, as long as I made a firm and steadfast resolution never to fail to observe them.[9]

The first was never to accept anything as true if I did not know clearly that it was so; that is, carefully to avoid prejudice and jumping to conclusions, and to include nothing in my judgements apart from whatever appeared so clearly and distinctly to my mind that I had no opportunity to cast doubt on it.

The second was to subdivide each of the problems that I was about to examine into as many parts as would be possible and necessary to resolve them better.

The third was to guide my thoughts in an orderly way by beginning with the objects that are the simplest and easiest to know and to rise gradually, as if by steps, to knowledge of the most complex, and even by assuming an order among objects in cases where there is no natural order among them.

And the final rule was: in all cases, to make such comprehensive enumerations and such general reviews that I was certain not to omit anything.

The long chains of inferences, all of them simple and easy, that geometers normally use to construct their most difficult demonstrations had given me an opportunity to think that all the things that can fall within the scope of human knowledge follow from each other in a similar way, and that as long as one avoids accepting something as true which is not so, and as long as one always observes the order required to deduce them from each other, there cannot be anything so remote that it cannot eventually be reached nor anything so hidden that it cannot be uncovered. I did not have great difficulty in deciding where one should begin, for I already knew to begin with what was simplest and easiest to know. And when I considered that, among all those who had previously searched for truth in the sciences, mathematicians were the only ones who were able to find some demonstrations, i.e. inferences which were certain and evident, I had no doubt that I ought to begin

[PART TWO]

with the same things that they had examined, even if I expected nothing else from this except that they would accustom the mind to being nourished on truths and to not being satisfied with faulty reasoning. In doing this it was not my intention, however, to try to learn all the special sciences that are usually called 'mathematics',[10] for I saw that, despite the diversity of the objects involved, they all agree in considering only the various relations or proportions between the objects in question. Thus I thought that it would be best to examine these proportions in general, and to consider them only in contexts that would make it easier for me to know them. However I also did not limit them in any way to those contexts, so that I could subsequently apply them much better to all the others to which they would be applicable. Then, once I realized that in order to know them I would sometimes need to consider them separately, and sometimes simply to keep them in mind or to understand a number of them together, I thought that, in order to examine them better separately, I ought to consider proportions between lines, because I found nothing simpler nor anything that I could represent more distinctly to my imagination and my senses. But in order to keep them in mind or understand a number of them together, I would have to explain them by using some symbols, the briefest ones possible, and in that way I would borrow everything that was best in geometrical analysis and algebra, and would use one to correct all the defects of the other.

Since I am bold enough, in fact, to claim that the exact observance of these few rules that I had chosen made it so easy for me to unravel all the problems to which these two sciences extend that, in the two or three months I spent studying them – since I began with what was most simple and general, and each truth that I discovered was a rule that would later serve to discover other truths – not only did I solve many problems that I had previously thought were very difficult, but it also seemed to me towards the end that, with respect to those that remained unsolved, I could determine the means by which and the extent to which they could be resolved. Perhaps I might not seem to you to be too vain about this if you consider that, since there is only one truth about each thing, whoever discovers it knows as much as it is possible to know about it and that, for example, a child who has been taught arithmetic and has done an addition in accordance with its rules, can be sure of having found everything that the human mind could find about the sum in question. For, after all, the method that teaches us to follow the correct order, and to enumerate precisely all the relevant facts in whatever we are looking for, contains everything that gives certainty to the rules of arithmetic.

But what satisfied me most about this method was that, by using it, I was assured of using my reason in everything – if not perfectly, at least as well as I was able to – and I felt that, by practising it, my mind became gradually more accustomed to conceiving of its objects more clearly and distinctly and that, not having restricted it to any particular matter, I promised myself to apply it to the problems of other sciences as successfully as I had to those of algebra. Not that I would thereby have dared immediately to take on the task of studying every problem that would arise, for that itself would have been contrary to the order prescribed by the method. But having realized that their principles should all be borrowed from philosophy – where I had so far found none that was certain – I thought that, before all else, I had to try to establish some principles in philosophy. Since that was the most important thing in the whole world and one in which prejudice and jumping to conclusions were most dangerous, I thought I should not try to complete it until I had reached a more mature age than twenty-three (as I was then), and until I had spent a long time preparing myself for it, in advance, by rooting out from my mind all the incorrect views I had previously accepted, gaining many experiences that would later serve as the subject matter of my reasoning, and practising constantly the method that I had prescribed for myself so as to improve more and more at it.

[PART THREE]

Now before beginning to rebuild the house where one lives, it is not enough to knock it down, to provide building materials and architects (or to practise architecture oneself), and to have the plans drawn carefully; it is also necessary to have provided oneself with another

[PART THREE]

house in which to live comfortably while the rebuilding is taking place. In a similar way, in order to avoid being indecisive about my actions during the interval when I would be forced by reason to be indecisive in making judgements, and to live as happily as possible during that time, I devised a provisional morality that included only three or four maxims, which I would like to share with you.

The first was to obey the laws and customs of my own country, holding firmly to the religion in which, by the grace of God, I had been instructed from my infancy, and guiding myself in everything else by the most moderate and least excessive views that are generally accepted in practice by the most sensible people among those with whom I was to live. For I had begun, at that time, to think of my own views as worthless because I wanted to re-examine all of them, and I was convinced that I could do no better than follow those of the most sensible people. Although there may have been some people among the Persians or the Chinese who were just as sensible as ourselves, it seemed to me that it would be most useful to be guided by those with whom I had to live and that, to discover what they really believed, I should pay more attention to what they did than to what they said – not only because, given the corruption of our morals, there are few people who are willing to express everything they believe, but also because many do not know what they themselves believe. For the act of thinking by which we believe something is different from the act by which we know what we believe, and one often occurs without the other.

Among the many views that are equally accepted, I would choose only the most moderate, both because they are always the most practicable and, probably, the best, since all excess tends to be bad; I also wished to stray less from the correct path, if I happened to make a mistake and choose one of two complete opposites, when the other was the one that should have been chosen. In particular, I classified as an excess any promise by which freedom is reduced – not that I disapproved of the laws which, as a remedy for the fickleness of human minds, allow people to make vows or contracts that must be fulfilled for the protection of commerce, when the objective is good (or even when the objective is merely indifferent), but because I saw nothing in the world that always remained in the same condition and, for my own part, I was planning to perfect my judgements more and more and not make them worse, and I thought it would be a major lapse of common sense if, because I approved of something at one time, I was still obliged to consider it good subsequently when, perhaps, it may have ceased to be good or I may have ceased to think it was so.

My second maxim was to be as firm and resolute as possible in my actions and to follow the most doubtful views, once I had decided to do so, just as steadfastly as if they were very certain, thereby imitating travellers who, when they find themselves lost in a forest, should not make the mistake of turning in one direction after another or, even less, of staying in the same place, but should always walk in one direction in as straight a line as possible and not change it for trivial reasons, even if initially it was only chance that determined them to choose it. For, in this way, if they do not arrive exactly where they wish, they will eventually arrive somewhere, and they will probably be better off there than in the middle of a forest. In a similar way, in everyday life we often have to act without delay and it is a very certain truth that, whenever we are unable to identify the most true opinions, we should follow the most probable,[11] and even when we do not notice any more probability in one than in another, we should still choose some of them and think about them subsequently no longer as doubtful, in so far as they are relevant to our practical life, but as very certain and very true, because the reason that made us choose them is such. This was able to free me, from then on, from all the regrets and remorse that usually disturb the consciences of weak and wavering minds who allow themselves fickly to begin doing things that they think are good but that they later think are evil.

My third maxim was to try always to overcome myself rather than fortune, to change my desires rather than the structure of the world and, in general, to get used to believing that there is nothing that is completely within our control apart from our thoughts. Thus when we have done our best with respect to external things, anything that is lacking in our success is, from our point of view, absolutely impossible. This alone seemed to me to be enough to prevent me in future from desiring anything that I had not acquired, and thus to make me content. For since our will moves to desire only things that our understanding represents to it as in some way possible, it is certain that, if we consider all external goods as equally beyond our power, we would have no more regret because we lack those that we seem to deserve by birth, when we are deprived of them through no fault of our own, than we have for not possessing the kingdoms of China or Mexico. Thus by making virtue of necessity, as they say, we would no more wish to be healthy when we are sick, or to be free when we are in prison, than we currently desire to have a body made from matter which is as durable as diamonds or to have wings like a bird for flying. But I admit that one needs long practice and a frequently repeated meditation to get used to seeing everything from this perspective, and I believe that this was principally the secret of those philosophers who were able, in earlier times, to withdraw from fortune's dominion, to despise suffering and poverty and to rival their gods in happiness.[12] For by constant reflection on the limits set for them by nature, they convinced themselves that there was nothing so completely in their power, apart from their own thoughts, that this alone was enough to prevent them from having any affection for other things; and their absolute control over their thoughts gave them reason to think that they were richer, more powerful, freer and happier than others who, however favoured by nature and fortune they might be, never have such control over everything they wish because they lack this philosophy.

Finally, as a conclusion for this morality, I decided to review the various occupations that are open to people in this life, and to try to pick the best one; and without wanting to say anything about those of others, I thought I could do no better than persevere in the very same occupation I already had, that is, using my whole life to develop my reason and making as much progress as I could in discovering the truth in accordance with the method that I had prescribed for myself. Since beginning to use this method I had experienced such great satisfaction that I thought it was impossible to experience a more pleasant or more innocent happiness in this life; and by using this method I discovered each day some truths which seemed to be rather important and to be generally unknown to other people, and the satisfaction that resulted from this filled my mind so much that nothing else compared with it. Moreover, the three maxims above were based only on my plan to continue to learn. For since God has given each of us a light for distinguishing what is true from what is false, I could not have believed that I should be content with the views of other people for a single moment if I had not planned to use my own judgement to examine them in due course, and I could not have avoided having scruples about their opinions if I had not hoped thereby not to miss any opportunity to find better ones if such were available. I would not have been able to limit my desire or to be satisfied if I had not followed a path by which, thinking I was certain to acquire all the knowledge of which I am capable, I also thought I could acquire all the genuine goods that would ever be within my grasp. For since our will cannot follow or flee from anything except in so far as our understanding represents the thing in question as good or evil, judging well is enough to do good, and judging as well as possible is enough to do one's best, that is, to acquire all the virtues and, with them, all the other goods that one is capable of acquiring. When someone is certain of this, they cannot fail to be satisfied.

Having convinced myself of these maxims and having set them to one side, together with the truths of the faith that have always been among my primary beliefs, I thought I could begin freely to rid myself of all my other views. Since I hoped to finish this task better in discussions with other people than by remaining shut up any longer in the stove-heated room in which I had had all these thoughts, I set off again to travel before winter was completely over. During the following nine years I did nothing other than wander around the world, trying to be a spectator rather than an actor in the dramas that unfold there. In each subject matter, I reflected particularly on what might make it doubtful or give us an occasion for making a mistake, and, meantime, I rooted out from my mind all the errors that could have slipped in. Not that I thereby imitated the sceptics who doubt only for the sake of doubting and pretend to be permanently undecided; on the contrary, my whole plan was designed only to convince myself and to reject the shifting ground and sand in order to find rock or clay. I think I succeeded reasonably well in this in so far as, attempting to uncover the falsehood or uncertainty of the propositions that I studied – not by feeble conjectures, but by using clear and certain arguments – I found none so doubtful that I did not always draw some rather certain conclusion from it, even if it was only that it contained nothing that was certain. And just as, when knocking down an old house, one usually keeps the demolition material to be used for building a new one; so likewise in destroying all those opinions of mine that I judged to be ill-founded, I made various observations and acquired many experiences that I have since used to build up opinions which are more certain. Moreover, I continued to practise the method I had prescribed for myself. For besides taking care generally to guide all my thoughts in keeping with its rules, I set aside some hours from time to time that I used specifically to apply this method to mathematical problems, or even to some others that I could almost convert into mathematical problems by detaching them from all the principles of the other sciences which I found were not as secure, as you will see I have done with some of those discussed in this book.[13] Thus without apparently living differently from those who are concerned only to lead an agreeable and innocent life, who are careful to separate pleasures from vices, and who engage in any honourable pastime in order to enjoy their leisure without getting bored, I continued to follow my plan and to progress in knowledge of the truth, perhaps more than if I had merely read books or spent my time in the company of the learned.

Those nine years passed by, however, before I had made up my mind about the questions that are usually debated among educated people or had begun to look for foundations for a philosophy that would be more certain than what is generally accepted. The example of many excellent minds who had previously had this plan, but who seemed to me not to have succeeded, made me imagine such great obstacles that I might not have dared to embark on it so soon if I had not seen that some people were already spreading the rumour that I had finished the task. I could not say on what basis they came to that conclusion; if I contributed anything to it by what I have said in public, it must have been by admitting what I did not know more frankly than is customary for those who have done a little study and, perhaps, by making known the reasons I had for doubting many things that others thought were certain, rather than by boasting that I had something to teach. But since I had a strong desire not to want to be taken for anything other than what I was, I thought I should try by every possible means to become worthy of the reputation I had acquired; and it is exactly eight years since this desire made me resolve to move away from all the places where I had acquaintances and to retire here[14] to a country where the long duration of the war has resulted in a situation in which the armies involved in it serve only to make the fruits of peace available with much greater security and where, among a great crowd of busy people, who are more concerned with their own business than inquisitive about that of others, I have been able to live as solitary and withdrawn a life as in the most remote deserts, without lacking any of the conveniences that are available in the busiest towns.

[PART FOUR]

I do not know if I should tell you about the first meditations that I did there, because they are so metaphysical and unusual that they may not be to everyone's taste.[15] However, I find that I am in some way forced to talk about them, so that readers may judge whether the foundations I adopted are solid enough. I had realized for a long time that it is sometimes necessary, in our conduct, to act on the basis of opinions that are known to be uncertain as if they were indubitable, as already mentioned above. But since, at that time, I wanted to focus exclusively on the search for truth, I thought it was necessary to do the exact opposite, and to reject as absolutely false everything in which I could imagine the slightest doubt and to see, as a result, if anything remained among my beliefs that was completely indubitable. Thus, because our senses sometimes deceive us, I decided to assume that nothing was the way the senses made us imagine it. And since there are some people who make mistakes in reasoning and commit logical fallacies, even in the simplest geometrical proofs, and since I thought that I was as subject to mistakes as anyone else, I rejected as false all the arguments that I had previously accepted as demonstrations. Finally, since I thought that we could have all the same thoughts, while asleep, as we have while we are awake, although none of them is true at that time, I decided to pretend that nothing that ever entered my mind was any

[PART FOUR]

more true than the illusions of my dreams. But I noticed, immediately afterwards, that while I thus wished to think that everything was false, it was necessarily the case that I, who was thinking this, was something. When I noticed that this truth 'I think, therefore I am' was so firm and certain that all the most extravagant assumptions of the sceptics were unable to shake it, I judged that I could accept it without scruple as the first principle of the philosophy for which I was searching.

Then, when I was examining what I was, I realized that I could pretend that I had no body, and that there was no world nor any place in which I was present, but I could not pretend in the same way that I did not exist. On the contrary, from the very fact that I was thinking of doubting the truth of other things, it followed very evidently and very certainly that I existed; whereas if I merely ceased to think, even if all the rest of what I had ever imagined were true, I would have no reason to believe that I existed. I knew from this that I was a substance, the whole essence or nature of which was to think and which, in order to exist, has no need of any place and does not depend on anything material. Thus this self – that is, the soul by which I am what I am – is completely distinct from the body and is even easier to know than it, and even if the body did not exist the soul would still be everything that it is.

After that, I thought about what a proposition generally needs in order to be true and certain because, since I had just found one that I knew was such, I thought I should also know what this certainty consists in. Having noticed that there is nothing at all in the proposition 'I think, therefore I am' which convinces me that I speak the truth, apart from the fact that I see very clearly that one has to exist in order to think, I judged that I could adopt as a general rule that those things that we conceive very clearly and distinctly are all true. The only outstanding difficulty is in recognizing which ones we conceive distinctly.

Then, by reflecting on the fact that I doubted and that, consequently, my being was not completely perfect – for I saw clearly that it was a greater perfection to know than to doubt – I decided to find out where I learned to think about something that was more perfect than myself, and I knew clearly that this had to be from some nature that was in fact more perfect. As regards the ideas I had of many other external things, such as the sky, the earth, light, heat and thousands of others, I did not have any comparable difficulty in knowing where they came from, because I did not notice anything in them that seemed to make them superior to me and therefore I was able to believe that, if they were true, they depended on my nature to the extent that they contained any perfection and, if they were not true, I got them from nothing, that is, they were present in me because of some deficiency in me. But the same would not apply to the idea of a being that was more perfect than me. To get such an idea from nothing was something manifestly impossible. And I could not have received it from myself either, because it was just as impossible for something that is more perfect to result from and depend on something less perfect as for something to proceed from nothing. Thus the only remaining option was that this idea was put in me by a nature that was really more perfect than I was, one that even had in itself all the perfections of which I could have some idea, that is – to express myself in a single word – by God.

I would add that, because I knew some perfections that I did not have myself, I was not the only being in existence (with your permission, I shall freely use scholastic terminology here), but that it necessarily had to be the case that there was some more perfect being on which I depended and from which I received everything that I had. For if I had been alone and independent of everything else, so that I myself was the source of the little perfection by which I participated in the perfect being, then by the same argument I would have been able to have, from myself, all the extra perfection that I knew I lacked, and thus I would myself be infinite, eternal, immutable, omniscient, omnipotent, and indeed have all the perfections that I was able to recognize in God. For according to the arguments I have just constructed, all I had to do in order to know the nature of God – in so far as my nature was capable of doing so – was to ask, about all the things of which I found some idea in myself, whether or not it is a perfection to possess them. I was convinced that none of them which involved some imperfection was in God, but that all the others were. Thus I saw that doubt, fickleness, sadness and similar things could not have been in him, since even I myself would have been very glad to be free from them. I also had ideas

[PART FOUR]

of many observable and bodily things; for even if I supposed that I was dreaming and that everything that I saw or imagined was false, I still could not deny that the ideas were truly in my thought. But because I already knew in myself very clearly that there is a distinction between intellectual and bodily natures, when I considered that every composition is an indication of dependence and that dependence is manifestly a defect, I concluded that it could not be a perfection in God to be composed of these two natures and that, consequently, he was not such. Instead, I concluded that if there were some bodies in the world, or some intellects or other natures, which were not completely perfect, their existence must therefore depend on his power in such a way that they could not subsist without him for a single moment.

I then tried to find other truths and, turning to the object studied by geometers – which I conceived of as a continuous body or space extended indefinitely in length, breadth and height or depth, divisible into different parts that could have various shapes and sizes and could be moved or transposed in every way, for geometers presuppose all that in the object of their study – I reviewed a sample of their simplest demonstrations. When I noticed that the great certainty that everyone attributes to these demonstrations is based only on the fact that they are conceived as evident, in accordance with the rule that I stated earlier, I also realized that there was nothing at all in them that would convince me of the existence of their object. For example, I saw clearly that, if we assume a triangle as given, its three angles would have to be equal to two right angles; however, for all that, I saw nothing that convinced me that there was any triangle in the world. In contrast, when I returned again to examine the idea I had of a perfect being, I found that existence was included in it in the same way as, or even more evidently than, the idea of a triangle includes its three angles being equal to two right angles or the idea of a sphere includes the equidistance from its centre of all its parts [on the surface] and that, consequently, it is at least as certain as any geometrical demonstration could ever be that God, who is this perfect being, is or exists.

But the reason why many people are convinced that there is some difficulty in knowing this, and even in knowing what their soul is, is that they never raise their minds above observable things, and they are so used to not thinking about anything without imagining it — which is an appropriate way of thinking about material things — that anything that is not imaginable seems to them to be unintelligible. This is clear enough from the fact that even philosophers in the schools adopt the maxim that there is nothing in the intellect that was not previously in the senses, although it is certain that the ideas of God and of the soul have never been there. It seems to me that those who wish to use their imagination to understand these ideas are doing exactly the same as if they wished to use their eyes to hear sounds or to smell odours, except for this difference: the sense of sight is no less convincing about the truth of its objects than smell or hearing are, whereas neither the imagination nor our senses could ever convince us of anything if our understanding did not intervene in the process.

Finally, if there are still some people who are not sufficiently convinced of the existence of God and of their soul by the arguments that I have presented, I want them to know that all the other things of which they think they are more convinced — such as having a body, or the existence of an earth or of the stars, and similar things — are less certain. For although we are morally certain[16] about these things, so that to doubt them would seem to be extravagant, it is also true that, when it is a question of metaphysical certainty, it cannot reasonably be denied that there is a basis for not being completely convinced about them once we realize that, while asleep, one can imagine having a different body in exactly the same way, or that one can see other stars and another earth where there is none. For how do we know that the thoughts that arise in us while we are dreaming are more false than others, since they are often no less vivid and explicit? Let the best minds study this question as much as they wish; I do not think they can provide a reason that is enough to resolve this doubt if they do not presuppose the existence of God. For, firstly, the very thing that I had accepted as a rule above — viz. that the things we conceive clearly and distinctly are all true — is guaranteed only because God is or exists, that he is a perfect being, and that everything in us derives from him. It follows from this that our ideas or notions, which are real things and come from God, cannot but be true to the extent that they are clear and distinct. Thus if we frequently have ideas that contain some falsity,

[PART FOUR]

that can arise only in the case of those that contain some confusion and obscurity because, to the extent that they do so, they participate in nothingness; in other words, they are present in us in this confused manner only because we are not completely perfect. It is evident that to claim that falsehood or imperfection as such result from God is no less contradictory than claiming that truth or perfection would result from nothingness. But if we did not know that any reality or truth in us comes from God, then no matter how clear and distinct our ideas might be, we would have no reason that would convince us that they had the perfection of being true.

But once knowledge of God and of the soul has thus made us certain of this rule, it is rather easy to know that the dreams we imagine while asleep should in no way make us doubt about the truth of the thoughts we have while awake. For if it happened, even while sleeping, that someone had a very distinct idea – for example, if a geometer discovered a new demonstration – the fact that they were asleep would not prevent it from being true. And as regards the most common mistake of our dreams, i.e. that they represent various objects to us in the same way as our external senses do, it is irrelevant that it provides an opportunity for us to doubt the truth of such ideas, because they can also mislead us rather frequently even when we are not asleep, as when those with jaundice see everything coloured yellow, or when stars and other very distant bodies appear much smaller to us than they really are. After all, whether we are awake or asleep, we should never allow ourselves to be convinced of anything except by the evidence of our reason; and it should be noted that I say 'our reason', and not 'our imagination' or 'our senses'. Although we see the sun very clearly, we should not for that reason judge that it is only as large as it seems; and we could easily imagine distinctively a lion's head attached to a goat's body without, for that reason, thinking that there is a chimera in the world, because reason does not tell us that what we see or imagine in this way is true. It tells us instead that all our ideas or notions must have some basis in truth, for otherwise it would be impossible for God, who is absolutely perfect and true, to put them in us. Since our reasoning is not as evident or complete while we are asleep as when we are awake, although what we imagine while asleep is sometimes as vivid and explicit, or even more so, reason also tells us that our thoughts cannot all be true, since we are not absolutely perfect, and that whatever truth our ideas possess should infallibly be found in those we have while awake rather than in our dreams.

[PART FIVE]

I would be very happy to continue and to reveal the whole chain of other truths that I deduced from these first ones. However, in order to do that, I would have to discuss here many questions that are disputed among the learned, and I do not wish to become entangled in those issues. I think it would be better, therefore, if I were to refrain from discussing them and if I simply identified them in a general way, so as to leave it to those who are wiser to judge whether it would be useful if the public were informed about them in more detail. I remain firmly committed to the decision I have made not to assume any other principle apart from the one I have just used to demonstrate the existence of God and of the soul, and not to accept anything as true that does not seem more clear and distinct to me than the proofs of geometers seemed to me in the past. Nevertheless, I dare to claim that not only have I found a way to satisfy myself in a relatively short time about all the principal difficulties that are usually discussed in philosophy, but I have also noticed certain laws which God has so established in nature and of which he has impressed such notions in our souls that, having reflected on them sufficiently, we could not doubt that they are observed exactly in everything that exists or that happens in the world. And in thinking about the consequences of these laws, I believe I have discovered more truths that are more useful and important than everything I had previously learned or had ever hoped to learn.

But since I tried to explain the principal truths among them in a treatise that I was prevented from publishing by a number of considerations, the best I can do to make them known is to provide a summary here of that treatise.[17] My plan was to include in it everything that I thought I knew, before I began writing it, about the nature of material

[PART FIVE]

things. But just as painters cannot represent equally well on a flat surface all the different sides of a solid body — by choosing one of the principal sides to face towards the daylight, they leave all the others in the shade and do not allow them to appear except in so far as one can see them by looking at the principal side — in the same way, I was afraid that I could not include in my discourse everything that I had thought about. Thus I undertook to explain in it reasonably fully only what I understood about light and, at the same time, to add something about the sun and the fixed stars, because most light results from them; about the skies, because they transmit the light; about the planets, comets and the earth, because they reflect it; and in particular about all the bodies that are on earth, because they are either coloured, transparent or luminous; and, finally, about human beings, because they are the perceivers of light. However, in order to camouflage all these things to some extent and to be able to say more freely what I thought about them without having either to accept or refute the views that are in vogue among the learned, I decided to leave this whole real world to their disputes and to speak only about what would happen in a new world if God were to create somewhere, in imaginary space, enough matter to compose such a world, and if he were to impart random and varying motions to the different parts of this matter so that it constituted a chaos as confused as poets could imagine and if, subsequently, he did nothing else except lend his ordinary cooperation to nature and allow it to act in accordance with the laws that he had established.[18]

Thus I first described this matter and tried to describe it in such a way that there is nothing in the world, it seems to me, which is more clear or more intelligible, apart from what was said earlier about God and the soul. I even assumed, explicitly, that it contained none of those forms or qualities about which they dispute in the schools nor, generally, anything which was not so naturally known to our souls that one could not even pretend to be ignorant of it. Moreover, I showed what the laws of nature are and, without basing my reasons on any other principle apart from the infinite perfections of God, I tried to demonstrate all the laws that may have seemed to be doubtful, and to show that they are such that, even if God had created many worlds, there could never be one in which these laws failed to be observed. Then I showed how the largest part of the matter in this chaos should dispose and arrange itself, according to these laws, in a certain way which would make it similar to our skies; how, meanwhile, some parts of this matter should compose an earth, some should compose planets and comets, and others should compose a sun and fixed stars. Then, expanding on the subject of light, I explained at considerable length what kind of light should be found in the sun and the stars, and how it traversed, in an instant, the immense spaces of the heavens and how it would be reflected from the planets and comets towards the earth.

I also added many things about the substance, position, movements and all the various qualities of these heavens and stars, so that I thought I had said enough to show that there is nothing observed in those of this world which ought not to, or, at least, which could not, appear in exactly the same way in those of the world I was describing. Then I came to speak specifically about the earth – how, although I had explicitly assumed that God had not put any weight in the matter of which it is composed, all its parts would necessarily tend precisely towards its centre; how, since there is water and air on its surface, the disposition of the skies and the stars, especially of the moon, should cause an ebb and flow on its surface that would be similar in every respect to what is observed in our seas; how it should also cause a certain current – of water, as much as of air – from the east to the west, similar to what is also observed in the tropics; how mountains, seas, fountains and rivers could naturally be formed there, and how metals could be present in mines and how plants could grow in the fields and, generally, how all the bodies that are called mixed or composed could be engendered there. Among other things, since there was nothing in the world that I knew of, besides the stars, which produced light except fire, I applied myself to explain very clearly everything that pertains to the nature of fire, how it occurs and how it is fed, and how it sometimes has only heat without light and sometimes only light without heat, how it can introduce different colours in various bodies and many other qualities, how it melts some bodies and hardens others, how it can consume almost all bodies or change them into ashes and smoke, and finally how from these ashes it forms glass simply by the force of its action. Since this transformation of ashes into glass seems to me to be as admirable as any other transformation that occurs in nature, I took special pleasure in describing it.

However, I did not wish to conclude from all these things that this earth was created in the way I had suggested, because it is much more probable that God made it, from the beginning, in the way it was to be.[19] But it is certain – a view that is widely accepted among theologians – that the action by which God conserves the world now is exactly the same as that by which he created it; so that even if, at the beginning, he had not given it any other form apart from that of a chaos, once he had established the laws of nature and had lent it his support to move as it usually does, one could believe, without denying the miracle of creation, that simply as a result of that, all things which are purely material would have been able, in time, to become as we observe them today. Their nature is also much easier to understand when one sees them developing gradually in this way than when one thinks of them only as being fully formed.

I moved on from the description of inanimate bodies and plants to that of animals and, in particular, to that of human beings. But since I did not yet have enough knowledge about this topic to speak in the same way as about the rest – i.e. by demonstrating effects by causes, and by showing from what seeds and in what way nature must produce human bodies – I was content to assume that God formed a human body entirely similar to one of ours, both in the external shape of its members and in the internal structure of its organs, without composing it of any other matter apart from what I have described above, and without putting into it, at the beginning, any rational soul or anything else that would function as a vegetative or sensitive soul. All he did was to stimulate in its heart one of those fires without light which I have already explained, and which I did not conceive as having any nature other than that which heats hay when it is harvested before it is dry or makes new wine bubble when it is left to ferment on the stems. In examining the functions that could occur in the body as a result of this heat, I found precisely all those that can occur in us without us thinking about them and, therefore, without our soul contributing to them – i.e. the part that is distinct from the body, about which it was said above that its nature is only to think; these are all the same functions in which one could think that animals without reason resemble us. However, I could not find any of those functions which, since they depend on thought, are the only ones that belong to us as human beings, whereas I found all of them there subsequently once I had assumed that God created a rational soul and joined it to this body in a certain way that I described.

But to show how I dealt with this subject matter, I would like to provide here an explanation of the movement of the heart and arteries. Since this is the first and the most general motion that is observed in animals, one will judge easily from it what one should think about all the others. And to make it easier to understand what I shall say about it, I would like those who are not trained in anatomy to take the trouble before reading this to have cut open in front of them the heart of a large animal with lungs, because the heart in all of them is similar enough to the human heart, and to have shown to them the two chambers or cavities which are there. First, that which is on the right side, with which two very large tubes communicate, viz. the vena cava, which is the principal receptacle of the blood and is like the trunk of a tree, of which all the other veins of the body are branches, and the arterial vein (which was poorly named thus, because in fact it is an artery) which originates in the heart and, when it comes out of the heart, divides into many branches that spread out everywhere in the lungs. Then, the ventricle on the left side with which two tubes communicate in the same way, which are as large as, or are larger than, the former ones. These are the venous artery (also badly named, because it is merely a vein), which comes from the lungs, where it is divided into many branches that are intertwined with those of the arterial vein and those of the conduit called the windpipe, through which air enters when breathing; and the great artery which, coming out of the heart, sends its branches throughout the whole body.

I would also like them to be shown carefully the eleven small valves, which, like so many little doors, open and close the four openings in these two cavities. There are three of them at the opening of the vena cava, where they are arranged so that they cannot in any way impede the blood it contains from flowing into the right ventricle of the heart,

[PART FIVE]

but they can prevent it completely from flowing out of it. There are three valves at the entrance to the arterial vein, which, since they are arranged in exactly the opposite way, easily allow the blood in this ventricle to pass into the lungs, but prevent the blood in the lungs from returning to the heart. There are also two more valves at the entrance to the venous artery that allow the blood to flow from the lungs towards the left cavity of the heart, but block its return; and there are three at the entrance of the large artery that allow blood to flow out of the heart but prevent its return. There is no need to look for any other explanation of the number of these valves apart from the fact that the opening of the venous artery is, because of its location, oval in shape and can easily be closed by two valves, whereas the others, which are round in shape, can be closed better by three of them. Moreover, I would like it if readers were asked to consider that the large artery and the two arterial veins are much harder and stronger in composition than the venous artery and the vena cava, and that the latter two become enlarged before entering the heart and are like two pouches there, called the auricles of the heart, which are composed of flesh similar to that of the heart. I would also like them to consider that there is more heat in the heart than in any other part of the body and, finally, that this heat can make a drop of blood which falls into its cavities expand and dilate suddenly, as all liquids usually do when they are made to fall, drop by drop, into a very hot vessel.

After that, I need say nothing else to explain the heart's motion except that, when its ventricles are not filled with blood, some blood flows necessarily from the vena cava into the right ventricle and from the venous artery into the left one, because these two vessels are filled with blood and their openings, which face towards the heart, cannot therefore be blocked. However, as soon as two drops of blood enter in this way, one into each ventricle, these drops – which must be very large, because the openings through which they enter are very large and the vessels from which they come are very full – are rarefied and expand because of the heat that they encounter there. In that way they cause the whole heart to expand, and push and close the five little doors that are at the entrances of the two vessels from which they came, thereby preventing any more blood from descending into the heart. While they continue to rarefy more and more, they push and open the other six small doors that are at the entrances of the other two vessels through which they exit, thereby causing all the branches of the arterial vein and of the large artery to expand at almost the same time as the heart itself. Almost immediately afterwards, the heart contracts, and so do the arteries, because the blood that has entered them becomes cooled there and their six small doors are closed, while the five doors of the vena cava and venous artery reopen, and allow in two more drops of blood which immediately cause the heart and arteries to expand, in exactly the same way as the previous drops. Since the blood that enters the heart in this way passes through these two pouches which are called its auricles, it follows that their movement is the opposite to that of the heart, and that they contract when it expands.

Finally, in order that those who are not familiar with the force of mathematical demonstrations and are unaccustomed to distinguishing true reasons from probable ones, may not be tempted to reject this without examination, I wish to advise them that this movement that I have just explained follows just as necessarily from the disposition of the organs alone (which can be observed in the heart with the naked eye), from the heat (which one can feel there with one's fingers), and from the nature of the blood (that can be known by experience), as the motion of a clock follows from the force, arrangement and shape of its counterweights and wheels.

But if one asks how the blood in the veins is not exhausted by flowing continuously into the heart, and how the arteries are not filled by it too much because all the blood that flows through the heart goes into them, I need not say anything else in reply except what has already been written by an English physician,[20] who deserves to be praised for having broken the ice on this issue and for having been the first who taught that there are numerous small passages at the extremities of the arteries through which the blood that they receive from the heart enters into the small branches of the veins, from which it returns immediately to the heart, so that its motion is nothing but a constant circulation. He proves this very well from the common experience of surgeons who, when they ligate the arm moderately tightly above the

[PART FIVE]

place where they open a vein, cause the blood to come out more abundantly than if they had not ligated it. The very opposite would happen if they ligated lower down, between the hand and the opening, or if they ligated very tightly above the opening. For it is clear that a moderately tight ligature can impede the blood that is already in the arm from returning to the heart through the veins; but it cannot thereby prevent new blood arriving from the arteries, because they are located below the veins and their walls are harder and are therefore less easy to press, and because the blood that comes from the heart tends to move with more force through them towards the hand than to return from there towards the heart through the veins. Since this blood comes out of the arm through the incision in one of the veins, there must necessarily be some passages below the ligature, i.e. towards the end of the arm, through which it can flow there through the arteries. He also proves equally well what he says about the way the blood flows through certain small membranes, that are so arranged at different places along the veins that they do not allow the blood to pass from the centre of the body towards its extremities, but only to return from the extremities towards the heart. He also proves it from the experience that shows that all the blood in the body can leave the body in a very short time through a single artery if it is cut, even if it had been very tightly ligated near the heart and if it were cut between the heart and the ligature, so that one has no reason to think that the blood which would emerge comes from anywhere else except the heart.

But there are many other things that show that the true cause of this motion of the blood is the one I indicated. For example, firstly, the difference that one can observe between the blood that comes from the veins and that which comes out of the arteries can result only from the fact that, when it is rarefied and as it were distilled in passing through the heart, it is thinner, livelier and warmer immediately after coming out – that is, when it is in the arteries – than it is a little while before it enters the heart, i.e. when it is in the veins. If one observes it, one will find that this difference appears clearly only near the heart, and does not appear as much in parts of the body that are far from it. Secondly, the hardness of the membranes, of which the arterial vein and the great artery are composed, is enough to show that the blood strikes against them with greater force than against those of the veins. And why would the left cavity of the heart and the large artery be larger and wider than the right cavity and the arterial vein, if it were not that the blood from the venous artery – which, since it passed through the heart, has been only in the lungs – is thinner and becomes rarefied more easily and to a greater extent than that which comes directly from the vena cava? And what can physicians discover in feeling the pulse if they do not know that, as the blood changes its nature, it can be rarefied by the heat of the heart more or less strongly and more or less quickly than before? And if one examines how this heat is passed on to other parts of the body, must one not concede that it happens by means of the blood which, in passing through the heart, is reheated and spreads from there throughout the whole body?

It follows from this that, if one removes the blood from some part of the body, one thereby also removes the heat from it; and even if the heart were as hot as glowing iron, it would not be enough to reheat the hands or the feet as much as it actually does, if it did not continually send new blood there. This also shows that the true function of respiration is to bring enough fresh air into the lungs so that the blood, which comes to the lungs from the right cavity of the heart, where it was rarefied and, as it were, changed into vapours, thickens there and is immediately converted into blood before falling again into the left cavity, without which it would not be able to serve as nourishment for the fire that it encounters there. This is confirmed by observing that animals which have no lungs also have only one cavity in the heart, and that children who cannot use their lungs while they are enclosed in their mothers' wombs have an aperture through which blood from the vena cava flows into the left cavity of the heart, and a tube through which it comes from the arterial vein into the large artery without passing through the lung. Again, how would digestion take place in the stomach, if the heart did not send it some heat through the arteries and also some of the more fluid parts of the blood that assist in breaking down the foods that have been put there? Is it not easy to understand the action that converts the juice of these foods into blood, if one considers that it is distilled as it passes repeatedly through the heart, perhaps more than one or two hundred times per

[PART FIVE]

day? And what more is required to explain nutrition, and the production of various humours that are found in the body, except to say that the force with which the blood passes from the heart to the extremities of the arteries when it is rarefied causes some of its parts to come to rest among those of the various members through which they flow, and they displace some of those that they expel; and, depending on the position, shape or small size of the pores they encounter, some of them go to some parts rather than to others in the same way that everyone may have seen that different sieves, with holes of varying sizes, are used to separate different grains from one another.

Finally, what is most noticeable in all this is the generation of animal spirits; they are like a very subtle wind or, rather, like a very pure and lively flame which, rising continually in great abundance from the heart to the brain, move from there through the nerves into the muscles and provide motion to all parts of the body. There is no need to imagine any other cause that makes the most penetrating and fast-moving parts of the blood the most suitable for constituting these spirits, and makes them move towards the brain rather than elsewhere, apart from the fact that the arteries, which carry them there, are those that come from the heart in a more direct line than any other arteries; according to the rules of mechanics – which are the same as the laws of nature – when many things tend together to move in a certain direction and there is not enough room for all of them, in the way in which particles of blood that exit from the left cavity of the heart tend towards the brain, the weaker and slower among them must be turned aside by the stronger particles which, in this way, travel there on their own.

I had explained all these things in sufficient detail in the treatise that I had planned to publish earlier.[21] Then I had shown, in the same place, what the structure of the nerves and muscles of the human body would have to be in order for the animal spirits in the body to have the power to move its members, as one sees when heads, soon after they have been cut off, still move and bite the ground even though they are no longer alive; what changes must be made in the brain to cause waking, sleep and dreams; how light, sounds, odours, tastes, warmth and all the other qualities of external objects can impress different ideas on it through the senses; how hunger, thirst, and the other internal passions can also send their ideas there; what part of the brain should be taken as 'the common sense', where these ideas are received;[22] what should be taken as the memory, which stores the ideas, and as the imagination, which can vary them in different ways and compose new ones and, by the same means, distribute the animal spirits to the muscles and cause the limbs of the body to move in as many different ways as our own bodies can move without the will directing them, depending on the objects that are present to the senses and the internal passions in the body.

This will not seem strange to those who know how many different automota or moving machines can be devised by human ingenuity, by using only very few pieces in comparison with the larger number of bones, muscles and nerves, arteries, veins and all the other parts in the body of every animal. They will think of this body like a machine which, having been made by the hand of God, is incomparably better structured than any machine that could be invented by human beings, and contains many more admirable movements. I specifically paused to show that, if there were such machines with the organs and shape of a monkey or of some other non-rational animal, we would have no way of discovering that they are not the same as these animals. But if there were machines that resembled our bodies and if they imitated our actions as much as is morally possible, we would always have two very certain means for recognizing that, none the less, they are not genuinely human. The first is that they would never be able to use speech, or other signs composed by themselves, as we do to express our thoughts to others. For one could easily conceive of a machine that is made in such a way that it utters words, and even that it would utter some words in response to physical actions that cause a change in its organs – for example, if someone touched it in a particular place, it would ask what one wishes to say to it, or if it were touched somewhere else, it would cry out that it was being hurt, and so on. But it could not arrange words in different ways to reply to the meaning of everything that is said in its presence, as even the most unintelligent human beings can do. The second means is that, even if they did many things as well as or, possibly, better than any one of us, they would infallibly fail in others. Thus one would discover that they did not act on the basis of knowledge, but merely as a result of the disposition of their organs. For whereas reason is a universal instrument that can be used in all kinds of situations, these organs need a specific disposition for every particular action. It follows that it is morally impossible for a machine to have enough different dispositions to make it act in every human situation in the same way as our reason makes us act.

Now one could also recognize the difference between human beings and beasts by these same two means. For it is very noticeable that there are no human beings so unintelligent and stupid, including even mad people, who are incapable of arranging different words and composing from them an utterance by which they make their thoughts understood; whereas there is no other animal, no matter how perfectly and favourably born it may be, which acts similarly. This does not result from the fact that they lack organs, for one sees that magpies and parrots can utter words as we do, but they still cannot speak as we do, that is, by showing that they think what they say; whereas human beings who are born deaf and dumb, and are thus deprived, as much as or even more than beasts, of the organs that are used by other people to speak, are accustomed to inventing some signs themselves by which they make themselves understood to those who are usually in their company and have the time to learn their language. And this shows not only that beasts have less reason than human beings, but that they have none at all. For it is clear that, to know how to speak, very little reason is required. And since one notices that there is as much inequality between animals of the same species as there is among human beings, and since some of them are easier to train than others, it is not credible that the most perfect representatives of the monkey or the parrot species would not be equal to one of the most stupid children or, at least, a child with a defective brain, if the nature of their soul were not completely different from ours.

One should not confuse words with the natural movements that express passions and that can be imitated both by machines and by animals; nor should one think, as some of the ancients did, that beasts speak although we do not understand their language. For if that were true, since they have many organs that correspond to ours, they would be able to make themselves understood by us as much as by other animals. It is also something very remarkable that, although there are many animals that show more skill than us in some of their actions, one still notices that the same animals do not show any skill at all in many others. Thus whatever they do better than us does not prove that they have a mind because, on this assumption, they would have more intelligence than any of us and would be better at everything. It proves, rather, that they have no intelligence at all, and that it is nature which acts in them in accordance with the disposition of their organs, just as we see that a clock, which is made only of wheels and springs, can count the hours and measure time more accurately than we can with all our efforts.

After that, I had described the rational soul and had shown that it could not in any way be drawn from the potentiality of matter[23] as could the other things that I spoke about, but that it has to be specially created. I also showed how it is not enough if, with the possible exception of moving its limbs, it is lodged in the human body as a pilot in their ship, but that it has to be joined and united more closely with the body in order to have, in addition, sensations and appetites like ours and thus constitute a real human being. Finally, I discussed the soul at some length here because it is among the most important subjects; for, apart from the error of those who deny God – which I think I have adequately refuted above – there is none that more readily leads weak minds away from the straight path of virtue than to imagine that the soul of beasts has the same nature as ours and, consequently, that we have nothing to fear or hope for, after this life, any more than flies or ants. However, when we know how much these souls differ, we understand much better the reasons that prove that our soul is of such a nature that it is completely independent of the body, and therefore that it does not have to die with it. And since one can see no other causes that destroy the soul, one is naturally led to judge that it is immortal.

[PART SIX]

It is now three years since I reached the end of the treatise that contains all these things and began to revise it for submission to a publisher, when I noticed that some people whom I respect, and whose authority over my actions can hardly be less than that of my own reason over my thoughts, had censured a physical theory which had been published a little earlier by someone else.[24] I am not saying that I shared this view; but I would not have noticed anything about it, prior to their censure, that I could have imagined as prejudicial either to religion or the state or, consequently, that would have prevented me from writing the same if I had been convinced of it by reason. This made me fear that I might have been mistaken about one of my own views, despite the great care I had always taken not to accept any new beliefs unless I had very certain demonstrations for them, and not to write anything about them that could turn out to be detrimental to anyone. That was enough to make me change my earlier decision to publish them. For even though the reasons why I had made that decision previously were very convincing, my natural inclination – which had always made me hate the work of writing books – caused me to find immediately other reasons that were enough to excuse me from publishing. These reasons, both for and against, are such that it is not only in my own interest to state them here, but the public may also have an interest in knowing them.

I have never attributed great significance to what came from my own mind, and, as long as I gathered no other results from the method I used apart from the fact that I solved a number of problems that belong to the speculative sciences, or that I tried to regulate my own conduct by the reasons that I learned from it, I did not believe that I should write anything about it. For, as far as conduct is concerned, everyone trusts their own sense so much that one could find as many reformers as there are heads, if others were allowed to assume responsibility for changing conduct, apart from those whom God has established as sovereigns over their peoples, or to whom he has given enough grace and zeal to be prophets. And even though I was very satisfied with my speculations, I believed that others also had their own, and that they may prefer them to mine. However, I soon acquired some general notions about physics; and when I began to test them in relation to specific problems and noticed where they can lead and how much they differ from the principles that had been used up to now, I thought I could not keep them hidden without sinning greatly against the law that obliges us to realize, as much as we can, the general welfare of all people. For they made me see that it is possible to achieve knowledge which would be very useful for life and that, in place of the speculative philosophy that is taught in the Schools, it is possible to find a practical philosophy by which, knowing the force and actions of fire, water, air, the stars, the heavens and all the other bodies that surround us, as distinctly as we know the various crafts of our artisans, we would be able to use them in the same way for all the applications for which they are appropriate, and thereby make ourselves, as it were, the lords and masters of nature.

This is desirable not only for the discovery of an infinite number of devices that would enable us to enjoy, without any effort, the fruits of the earth and all the goods we find there, but also, especially, for the preservation of health which is undoubtedly the foremost good and the foundation for all the other goods of this life. For even the mind depends so much on temperament and the disposition of one's bodily organs that, if it is possible to find a way to make people generally more wise and more skilful than they have been in the past, I believe that we should look for it in medicine. It is true that medicine as it is currently practised contains little of much use; but without wishing to disparage it, I am certain that there is no one, even among those who practise medicine, who does not concede that everything known about it is almost nothing in comparison with what remains to be discovered, and that we could avoid many infirmities, both of mind and body, and perhaps even also the decline of old age, if we had enough knowledge of their causes and of all the remedies that nature has provided for us. Since I planned to devote all my life to searching for such indispensable scientific knowledge, and since I found a path that seemed to be such that, by following it, I would infallibly find such knowledge, if one were not impeded either by the brevity of life or by

[PART SIX]

a lack of experiences, I thought that there was no better remedy for these two impediments than to communicate truthfully to the public what little I had found, and to encourage good minds to try to make further progress by contributing – each according to their inclination and ability – to the experiments that would have to be done, and also by communicating to the public all the things they learn. Thus, if later people begin at the point that their predecessors had reached and thereby join together the lives and labours of many, we would make much more progress together than each person could ever make on their own.

I also noticed, about experiences,[25] that the more we advance in knowledge, the more necessary they become. At the beginning, it is better to use only those that appear spontaneously to our senses and could not be unknown to us as long as we reflect even the slightest on them, than to look for more unusual or contrived experiences. The reason for this is that such unusual experiences are often deceptive, as long as we still do not know the causes of the more common experiences, and the circumstances on which they depend are almost always so special and so minute that it is very difficult to notice them. But the order that I observed in this project was the following. First, I tried generally to find the principles or first causes of everything that exists or that could exist in the world, without for this purpose thinking of anything other than God alone who created the world, and without deriving them from any source apart from certain seeds of truth that are naturally in our souls. I then looked for the first and most usual effects that could be deduced from these causes. I thought that, in doing so, I found the heavens, the stars, an earth and even, on the earth, water, air, fire and minerals, and other similar things that are the most common and most simple of all and, therefore, the easiest to know. Then, when I wanted to proceed to more detailed effects, I encountered so many of them that I thought it was impossible for the human mind to distinguish the forms or species of bodies that are found on earth from infinitely many others that could be there if it had been God's will to place them there. Consequently, I thought it was impossible to exploit them for our project, except by moving to the causes through their effects and by making use of many particular

45

experiences. Then, reviewing in my mind all the objects that were ever presented to my senses, I would venture to say that I noticed nothing among them that I could not explain quite easily by means of the principles that I had discovered.

But I must also acknowledge that the power of nature is so extensive and so great, and these principles are so simple and general, that I hardly ever notice any particular effect about which I do not realize immediately that it can be deduced from these principles in a number of different ways, and my biggest difficulty is usually in identifying in which of these ways it depends on them. I know of no other way of doing this except by then looking for some experiences such that their occurrence is not the same if the effect should be explained in one rather than another of these ways. Moreover I have reached a stage at which, it seems to me, I can see well enough how I should do most of the experiments that can be used for this purpose. But I also see that they are such and are so numerous that neither my personal effort nor my financial resources – even if I had a thousand times more than I actually have – would ever be enough for all of them. Thus in future I would advance more or less in my knowledge of nature in proportion to my capacity to do more or fewer experiments. I planned to show this in the treatise I had written, and to show very clearly how useful it would be for the public if I required all those who have a general desire for human welfare – i.e. all those who are genuinely virtuous, rather than those who are reputed to be such or falsely appear as virtuous – both to communicate to me the experiments they have already done and to assist me in searching for those that still need to be done.

But, in the mean time, I have had other reasons that made me change my mind and believe that I should continue to write everything that I judged to be in some way significant as I discovered the truth about it, and that I should take as much care as if I planned to publish such writings. The reason is to have much more time to study them properly, as one undoubtedly always examines more closely what one believes should be seen by many others than what one does only for oneself, and because it has often happened that things that seemed true to me, when I first thought about them, seemed false to me when I wished to write them down. The other reason is to lose no opportunity to benefit

[PART SIX]

the public if I am able to do so; and if my writings have any value, those who have them after my death may use them as they think fit. But I thought I should never agree to their publication during my life, so that the opposition and controversies to which they might be subject, or even the reputation, such as it is, that they could earn for me, would not give me an opportunity to waste the time that I plan to devote to study. For although it is true that everyone is obliged, as much as they can, to procure the well-being of others, and although not being useful to anyone is really worthless, still it is also true that our solicitude should extend beyond the present, and it is good to omit things that may possibly bring some benefit to those who are alive when planning to do others which could benefit posterity even more. I would also like it to be known that the little I have learned so far is almost nothing in comparison with what I do not know and do not despair of learning. For those who learn the truth in the sciences in a piecemeal way are almost the same as those who, when they begin to become rich, are less troubled when making large purchases than they had been previously, when they were poor, in making much smaller ones. Or they may be compared with army commanders whose forces usually increase in proportion to their victories; they need more skill to maintain their position after a loss in battle than they do, after winning one, to conquer towns and provinces. To try to overcome all the difficulties and errors that prevent us from reaching the truth in the sciences is really to become involved in battle; it is equivalent to losing a battle if we accept some false opinion about something rather general and significant, and it requires much more skill subsequently to recover the state in which we were previously, than to make great progress when one already has principles that are certain. For my part, if I have already found some truths in the sciences (and I hope that the contents of this volume will make people judge that I have found some[26]) I can say that they are merely the consequences and results of five or six principal difficulties that I overcame, and that I think about these as so many battles in which I had fortune on my side. I would even dare to say that I think I need to win only two or three other similar battles to realize my plans completely, and that I am not so old that, in the ordinary course of nature, I could not still have enough free time to do that. But I think I ought to manage the time that remains to me all the better in proportion to my hope of being able to use it well, and I would surely have many opportunities for wasting it if I published the foundations of my physics. For although they are almost all so evident that one needs only to understand them in order to believe them, and there is none for which I do not think I could provide demonstrations, nevertheless, since it is impossible for them to agree with all the different views of other people, I anticipate that I would often be distracted by the objections they would provoke.

One could argue that these objections would be useful both to make me realize my mistakes and, if I had anything worth while to reply, to provide others thereby with an improved understanding; and since many people can see more than an individual on their own, if they began immediately to make use of what I have discovered, they might also assist me with their discoveries. However, although I acknowledge that I am extremely subject to error and that I hardly ever trust the first thoughts that occur to me, still my experience of the objections that can be raised against me prevents me from hoping to benefit in any way from them. I have already often experienced the judgements of those whom I regarded as my friends, of others to whom I thought I was indifferent, and even of some about whom I knew that malice and envy would encourage them to discover things that affection concealed from my friends. But it has seldom happened that someone raised an objection to me that I had not completely anticipated, unless it was very far from being relevant. Thus I have almost never met any critic of my views who did not seem to be either less rigorous or less impartial than myself. Nor have I ever noticed that any truth was discovered that had been previously unknown, by means of the disputations that are practised in the Schools; while each side tries to win, they are much more preoccupied with making what is probable win than in weighing up competing arguments. And those who have been lawyers for a long time do not, as a result, become better judges afterwards.

The benefit that others would reap from publication of my thoughts could not be very great, because they are not yet developed enough not to need much more added before applying them in practice. I think

[PART SIX]

I can say, without vanity, that if anyone is capable of doing this, it ought to be me rather than someone else — not that there cannot be many minds in the world that are incomparably better than mine, but because something cannot be conceived and made one's own when it is learned from someone else as well as when one discovers it oneself. This is especially true in the present context, because I have often explained some of my views to people with very good minds and, while I was speaking to them, they seemed to understand them very clearly, but when they repeated them, I noticed that they almost always modified them in such a way that I could no longer accept them as my own.

At this point I want to plead here with future generations never to believe something if they are told that it originated from me, when I have not published it myself. I am not at all surprised at the extraordinary things attributed to all the ancient philosophers whose writings we do not possess; and for that reason I do not judge that their thoughts were very unreasonable, since they were among the best minds of their era, but that they have been reported inaccurately. Likewise we see that it has hardly ever happened that any of their followers surpassed them, and I am confident that the most dedicated among those who follow Aristotle today would think they were lucky if they had as much knowledge of nature as he had, even on condition that they would never come to have more in future. They are like the ivy that tends not to climb higher than the trees that support it, and often turns back down again after it has reached the top of the tree. For it seems to me that people also go back down again — that is, they in some way make themselves less wise than if they had abstained from study — who, not content with knowing everything that is intelligibly explained in their author, wish to find in them, over and above that, the solution to other problems about which the author says nothing and about which they may never have thought. However, this style of philosophizing is very appropriate for those who have only mediocre minds; for the obscurity of the distinctions and of the principles that they use is the reason why they can talk about everything as confidently as if they knew about it, and defend everything they say about it against the most subtle and knowledgeable, without leaving any room to convince them of their mistake. In doing this they seem to me to resemble a blind person who, in order to fight without any disadvantage against a sighted person, would bring them into the depths of a very dark cellar.

I can say that such people have an interest in my not publishing the principles of philosophy that I use. For, since they are very simple and very certain, by publishing them I would be doing almost the same as if I opened some windows and allowed light into the cellar into which they have descended for their fight. Even the best minds do not wish to know these principles; for if they wish to talk about everything and to gain a reputation for being learned, they will achieve that more easily by being content with what is probable, which can be found without great difficulty in all kinds of subjects, than in searching for the truth which can only be found, bit by bit, in some cases and which, when it comes to speaking about others, forces us to admit frankly that we are ignorant. If they prefer knowledge of a few truths to the vanity of appearing ignorant of none (and it is undoubtedly very preferable), and if they wish to follow a plan similar to mine, all they need is that I tell them nothing more than what I have already said in this *Discourse*. For if they are capable of going further than I have, they will also be all the more capable of discovering for themselves everything that I think I have discovered. Since all my investigations have proceeded according to a definite order, it is certain that what still remains for me to discover is, by its nature, more difficult and more concealed than what I have so far been able to find, and they would be less happy to learn it from me than to discover it themselves. Besides, the habit they will acquire by searching initially for easy things and moving gradually to more difficult ones will be more useful to them than all my instructions. Just as, for my own part, I am convinced that if I had been taught from my youth all the truths whose demonstrations I have been searching for since then, and if I had learned them without effort, I might never have known any others; at least, I would never have acquired the habit and facility that I think I have for always finding new truths as I apply myself to search for them. In a word, if there is one project in the world that cannot be completed as well by anyone else, apart from whoever started it, it is the one on which I am working.

It is true that, as regards experiments that can assist this project, they cannot all be done by one person working alone. But neither could

[PART SIX]

other hands apart from their own be usefully employed, except those of artisans or similar people who could be paid and for whom the hope of gain – which is very efficacious – would make them do precisely everything they are asked to do. Volunteers who might offer to assist, perhaps because of curiosity or a desire to learn – apart from the fact that they usually promise more than they deliver and make wonderful proposals, none of which ever succeeds – would inevitably expect to be compensated either by having various difficulties explained to them or, at least, by compliments and useless conversations that could not be provided without wasting time. As regards experiments that others have already done, even if they were willing to share them – something that would never be done by those who call them 'secrets' – for the most part they involve so many conditions and redundant details that it would be very difficult to decipher the truth from them. Besides, they would almost all be found to be so poorly explained or even so false, because those who did them tried to make them appear consistent with their principles, that, even if a few of them could be used, they would not be worth the time required to select them. Thus if there were someone in the world who was known with certainty to be capable of discovering the most important things, and those that are most useful for the common good, and if, for that reason, other people tried by every means to assist them to realize their plans, I cannot see that others could assist in any way except by providing financial assistance for the experiments that would be required and, otherwise, by preventing their free time from being wasted by the importunate demands of anyone else. But I am not so presumptuous as to promise anything extraordinary, nor do I feed myself on thoughts so vain that I imagine that the public should be very interested in my plans; nor do I have a soul so base that I would wish to accept from anyone at all any favour that I might be thought not to deserve.

All these considerations combined together were the reason why, three years ago, I was unwilling to publish the treatise that I had completed. I even decided not to publish during my lifetime any other treatise that was equally general in scope, or from which the foundations of my physics could be understood. Since then, however, two other reasons made me set down here some essays on particular issues and offer the public some account of my work and my plans. The first reason is that, if I failed to do so, certain people who knew about my earlier intention to publish some of my writings could imagine that the reasons why I refrained from doing so are more unfavourable to me than they really are. For, although I do not have an excessive desire for fame and, if I dare say so, I even detest it because I think it is incompatible with the peace that I esteem above everything else, I have also never tried to hide my actions as if they were crimes, nor have I taken great care to be anonymous. This was because I had thought I would harm myself and because it would have caused a certain disquiet that would not have been compatible with the perfect peace of mind that I sought. Also, since I always remained indifferent between trying to be known or unknown, I was unable to prevent myself from acquiring some kind of reputation, and I thought I should do my best at least to avoid getting a bad one. The second reason that made me write this is that I see every day more and more the delay involved in my project of self-instruction because of the infinity of experiences that I need, and because they are impossible for me to do without assistance from others. Although I do not flatter myself so much that I hope the public will share my interests greatly, at the same time I also do not want to fail myself so much that I provide for those who follow me an opportunity to reproach me some day that I could have left them many, much better things than I had done, if I had not been so negligent in explaining how they could contribute to my plans.

I also thought it would be easy to choose a few things which, since they were not subject to controversy and would not make me reveal my principles more than I wished, would still show clearly enough what I can and cannot achieve in the sciences. I cannot say if I have succeeded in this, and I do not want to anticipate anyone's judgement by speaking about my writings myself. But I would be very glad if they were examined and, to provide more opportunities for doing so, I ask all those who may have objections to them to take the trouble to send them to my publisher. Once the publisher tells me about them, I shall try to append my response at the same time and, in this way, readers will see both of them together, and will find it so much easier to make a judgement about the truth.[27] For I promise never to make long

[PART SIX]

replies, but simply to admit my mistakes very frankly if I know about them, or indeed, if I cannot recognize them, to say simply what I would think is required in order to defend what I have written, without adding an explanation of anything new, so as not to become involved endlessly in one topic after another.

If some of the issues that I have spoken about at the beginning of the *Dioptrics* and *Meteors* shock people initially, because I call them assumptions and seem not to want to prove them, they should have the patience to read the whole text attentively and I hope that they will be satisfied. For it seems to me that the arguments are interconnected in such a way that, as the last ones are demonstrated by the first which are their causes, the first arguments are demonstrated reciprocally by the last which are their effects. It should not be imagined that, by doing so, I commit the fallacy that logicians call a 'vicious circle'; since experience makes most of these effects very certain, the causes from which I deduce them are used not so much to prove as to explain them; but, in exactly the opposite way, it is the former which are proved by the latter. And I only called them 'assumptions' to let it be known that I think I can deduce them from those first truths that I have explained above, and that I explicitly chose not to do deduce them, to prevent certain minds – who imagine they know in one day everything that someone else had thought about for twenty years, as soon as they have been told two or three words about it and who, the more penetrating and lively they are, the more likely they are to be mistaken and the less capable of the truth – from availing themselves of the opportunity to build some extraordinary philosophy on what they would think are my principles and then to hold me responsible for it. For, as regards the views that are entirely my own, I do not excuse them as novel because, if one considers well the reasons that support them, I am confident that they will be found to be so simple and so consistent with common sense that they would seem to be less extraordinary and less strange than any other views one could have about the same subject matter. Nor do I boast about being the first person to discover any of them; rather, I never accepted any of them either because they were or were not expressed by others, but only because reason convinced me of them.

If artisans cannot implement immediately the invention I explained in the *Dioptrics*, I do not think that, for that reason, it can be said to be defective.[28] Since skill and practice are required to construct and adjust the machines that I described, even though no detail is omitted, I would be just as surprised if they succeeded on their first attempt as if someone were able to learn to play the lute very well in a single day, when they are provided with only a good tablature. And if I have written in French, which is my native language, rather than in Latin, which is the language of my teachers, it is because I hope that those who use only their pure natural reason will be better judges of my views than those who trust only ancient books. For those who combine common sense and study – and I hope that they alone will be my judges – will not be so partial to Latin, I am sure, that they refuse to listen to my arguments because I explain them in the vernacular.

In conclusion, I do not wish to speak here in detail about the progress in the sciences that I hope to make in the future, nor to be under an obligation to the public because of some promise that I am not certain to keep. I would say only that I decided to use the time that remains to me in life for nothing else except trying to acquire a knowledge of nature, from which one could draw some more reliable rules for medicine than those we have had up to now. Moreover, I am so strongly inclined to avoid all other kinds of project, especially those that can be helpful to some people only by harming others, that, if circumstances forced me to become involved in them, I do not believe that I could succeed. I declare this publicly here; I realize very well that it cannot help to make me important in the world, but I also have no wish to be such. And I shall always be more obliged to those by whose goodwill I would enjoy my leisure without disturbance, than to those who would offer me the most honourable employment in the world.

The End

Selected Correspondence, 1636–9

Descartes to Mersenne, March 1636[1]
(AT I, 338-40)

The reason I postponed replying to you was that I hoped to be able to tell you soon that I was busy with the publisher. That is why I came to this town [i.e. Leiden]; but although the Elseviers had previously indicated that they were very anxious to be my publishers, once they saw me here they imagined, I think, that I would not escape from them and they began to get difficult. That is why I have decided to leave them;[2] and although I could find a number of other publishers here, I shall none the less arrange nothing with any of them until I have heard from you, once there is no unreasonable delay involved. If you think my writings could be published in Paris more easily than here, and that you would be able to make the arrangements, as you were once kind enough to offer, I could send them to you as soon as I hear from you. The only problem is that my copy is not written any better than this letter, the handwriting and punctuation are just as poor, and the illustrations are drawn only by myself – in other words, very poorly. The result is that, if you did not interpret them for the engravers, they would not be able to understand them. Otherwise, I would be very happy if the whole thing were printed in a very good font, on very good-quality paper, and if the printer could supply me with at least two hundred copies, because I want to distribute them to many people. To let you know what I would like to have published: it will include four treatises, all in French, and the general title will be: *The Project of a Universal Science which could elevate our nature to its highest level of perfection. In addition, the Dioptrics, Meteors and Geometry, where, as illustrations of the universal science which it proposes, the most interesting things that the author could select are explained in such a way that even those who have not studied are able to understand them.* In this *Project* I unveil a part of my method, I try to prove the existence of God, and of the soul when it is separated from the body, and I add a number of other things which, I believe, will not be disagreeable to the reader . . .[3] Finally, in keeping with my earlier resolution, I do not wish to put my name to it and I ask you to say nothing about it to anyone – unless you think it appropriate to mention it to some publisher, to find out if they would wish to assist me, but without concluding a contract with them, please, until after you have heard from me.

Descartes to Constantijn Huygens, 27 February 1637[4]
(AT I, 620–21)

Mr Golius[5] informed me recently on your behalf that you thought the term 'Discourse' was superfluous in my title; that is one of the things that I have to thank you for. But I ask to be excused because I did not plan to explain my whole method, but merely to say something about it. I do not like to promise more than I can deliver and, for that reason, I wrote *Discourse on the Method*; in contrast, I wrote simply *The Dioptrics* and *The Meteors*, because I tried to include in them everything that belongs to my subject. If you are not satisfied with this explanation, and if you are kind enough to let me know your judgement, I shall follow it as if it were an inviolable law. I also think that I should remove the whole gloss that I had put at the end,[6] and leave simply these words: *Discourse on the Method, etc. and in addition, the Dioptrics, Meteors and Geometry, which are some tests of this method*. But I am afraid you may tell me that I behave as secretly as if I were a monk, if I continue to involve you with something which is so insignificant.

Descartes to Mersenne, March 1637
(AT I, 348–51)

I find that you have a very poor opinion of me, and that you think I am rather fickle and irresolute in my actions, if you think that I should consider your request that I change my plans and attach my first discourse to my Physics,[7] as if I should give it to the publisher the very same day that I see your letter! And I could not restrain myself from laughing, when I read the part where you say that I am making people kill me so that my writings can be seen sooner. I have no reply to this, except to say that my writings are already in a place and condition such that those who would kill me would never be able to find them, and if I do not die more in my own time and more content with those who are living, they will certainly not see them for more than a hundred years after my death.

I am very grateful to you for the objections you have written out for me, and I would ask you to continue to pass on to me all the objections you hear and to present them in the way that is most critical of me. That will be the best favour you could do me. For I am not in the habit of complaining when my wounds are being dressed, and those who do me the favour of instructing me and who teach me something will always find me very docile. However, I was unable to make much sense of your objection to the title. For I do not have *A Treatise on the Method* but *A Discourse on the Method*, which is the same as *A Preface or Notice about the Method*, to show that I do not plan to teach the method here but merely to speak about it. For – as can be seen from what I say about it – it consists in practice more than in theory, and I call the treatises which follow it *Tests of this Method*, because I claim that the things included in them could not have been discovered without it and that one can know from them how valuable it is, just as I have also inserted some metaphysics, physics and medicine in the first discourse to show that my method applies to all kinds of subject matter.

As regards your second objection, viz. that I have not explained at sufficient length how I know that the soul is a substance which is distinct from the body, that its nature is merely to think, and that this is the only thing which makes the demonstration of God's existence obscure; I concede that what you write about it is very true, and also that it makes my demonstration about the existence of God difficult to understand. But I was unable to discuss this question better without explaining more fully the falsehood or uncertainty which occurs in all judgements which depend on sensation and imagination, so as to show subsequently which judgements depend only on pure understanding, and how clear and certain they are. I omitted this very much on purpose – and for a number of reasons, primarily because I wrote in the vernacular – because I was afraid that feeble minds would first come to adopt enthusiastically the doubts and worries that I would have had to introduce, but would not be able subsequently to understand in the same way the reasons by which I would have tried to remove them, and thus I would have got them started on a bad journey without, possibly, being able to rescue them. But it is almost eight years since I wrote, in Latin, an introduction to metaphysics in which these issues are discussed at sufficient length, and if a Latin edition of my book is published, as is planned, I could have them included there.[8] Meanwhile I am convinced that those who examine carefully my arguments about God's existence will find that, the more trouble they take to look for mistakes in them, the more demonstrative they will find them, and I claim that, in themselves, they are clearer than any geometrical demonstrations. Thus it seems to me that they are obscure only to those who do not know how *to lead their minds away from the senses* [in Latin] in keeping with what I wrote on page 38.[9]

Descartes to an Unknown Correspondent [May 1637?]
(*AT I, 369–71*)

Although Father Mersenne did the exact opposite of what I had asked by revealing my name [as author], I still could not be angry with him because, by doing what he did, I am honoured to be known by someone as talented as you are. But I cannot agree with the proposal for a privilege[10] which, he informs me, he wishes to try to request for me; for he introduces me as praising myself and describing myself as the inventor of many great things, and he has me saying that I propose to provide the public with other treatises in addition to those that are already printed. This is contrary to what I wrote both at the beginning of page 77 [of the *Discourse*],[11] which serves as a preface, and elsewhere. But I am sure he will let you see what I am asking him, because I understand from the letter you have kindly written me that it was you who obliged me by suggesting to him some of the objections in response to which I wrote to him.

As regards the treatise on physics that you kindly request me to publish, I would not have been so imprudent in speaking about it as I did if I had not wished to see it published, on condition that the public want it and that I feel secure about it myself. But I want to tell you that the whole point of what I am publishing now is simply to prepare the way for it and test the water. For this reason I am proposing a general method that I am not actually teaching, but I try to provide tests of it by means of the three subsequent treatises that I attach to the discourse in which I speak about my method. The first has a mixture of philosophy and mathematics as its subject, the second one is devoted entirely to philosophy, and the third is exclusively mathematics.[12] In these treatises I can say that I did not refrain from speaking about anything (at least, about anything that can be known by the power of reasoning) because I believed that I did not know it. I therefore thought that this provided an opportunity to judge that I use a method by which I could explain all other subjects just as well, on condition that I had the experiences which would be necessary in each discipline and the time to reflect on them. Moreover, to show that this method applies to everything, I inserted something brief about metaphysics, physics and medicine in the first discourse. If I can get the world to see my method in this way, I would believe then that I no longer have any reason to fear that the principles of my physics would be received badly. And if I were to meet only judges who were as favourable as you are, I would have no fear about it in future.

Descartes to [Father Noël], October 1637[13]
(AT I, 454–6)

I am very glad to learn, from the letter you kindly wrote to me, that I am still lucky enough to have a place in your memory and affection. I also thank you for promising to have the book I had sent you examined by some of your colleagues who are most involved in such matters, and for obliging me so much by sending me their criticisms.[14] I would only wish, in addition, that you were willing to take the trouble to add your own comments to theirs; for I assure you that none of their criticisms would have as much authority with me as yours, and there are none to which I would defer more willingly than to yours. It is true that those of my friends who have already seen this book have told me that one needs time and study to be able to judge it properly, because one can be convinced fully of the initial parts (at least those of the *Dioptrics* and *Meteors*) only by knowledge of all the things which follow subsequently; and the things that follow cannot be understood well without a clear memory of all those which preceded them. That is why I would be especially indebted to you if you were willing to take the trouble to examine it or to arrange for others to do so. Since, in fact, my plan is to instruct only myself, those who would reproach me with some mistake would always please me more than those who praise me. Finally, there is no one who seems to me to have a greater interest in examining this book than those of your Society; for I already see so many people who are in the process of accepting what it contains that (especially in the case of the *Meteors*) I do not know how they could teach these subjects in future, as they do every year in most of your colleges, unless they either disprove or accept what I have written about them. And because I realize that the main reason why your confrères are very careful to reject every kind of novelty in philosophical matters is their fear that they would also cause some change in theology, I wish to advise you here, in particular, that there is nothing at all to fear in that respect from my writings; I have reason to be grateful to God that the views which, based on reflection on natural causes, seemed most plausible to me in physics have always been those which, among all of them, are most compatible with the mysteries of religion. I hope to show this clearly when I have an opportunity to do so.

Morin to Descartes, 22 February 1638[15]
(AT I, 537–40)

From the first time I had the honour of seeing and meeting you in Paris, I thought you had a mind that would be able to bequeath something rare and excellent to posterity. I am delighted to have seen my judgement vindicated by the wonderful book you have published on the subjects of mathematics and physics, which are also the two principal objects of my own scientific reflections. But just as, in the case of mathematics, you will have only people who admire the nobility of your mind, I do not think you will be surprised if, in the case of physics, there are some people who disagree with you. For, since you have kept to yourself knowledge of the principles and universal notions of your new physics (publication of which is ardently desired by all the learned), and since you have based your explanations only on analogies and hypotheses, about the truth of which people are at least in doubt, it would be sinning against the first rule of your method – which is a good one, and one that I am familiar with – to acquiesce in your explanations. And even though, according to page 76 of your *Method*,[16] experience makes most of the effects that you discuss very certain, nevertheless you know very well that the apparent movements of the heavens are deducible just as certainly from the hypothesis that the earth is stationary as from the hypothesis of its motion. Therefore, the experience of that particular appearance is not enough to prove which of these two causes is the correct one. If it is true that proving effects by means of an assumed cause and then proving the same cause by the same effects is not a vicious circle, then Aristotle misunderstood it, and one could say that it is impossible to commit this fallacy.

As regard the astronomers whom you propose to imitate on page 3 of the *Dioptrics*,[17] I shall not conceal my own view from you, which is as follows: anyone who does not formulate better hypotheses than what have been assumed to date by astronomers, will fare no better than they have in their consequences or conclusions, and may even do worse. For when they assume mistakenly the parallax of the sun, or the obliquity of the ecliptic, or the eccentricity of the apogee, the average motion or period of a planet, etc., far from drawing very true and very certain consequences from them, as you say on the page 3 mentioned, on the contrary they err subsequently in the motions or positions of the planets in proportion to the error of their false hypotheses, as is evident from the correspondence between their tables and the skies. And I think I was the first person in the world who, in my book on longitudes, provided astronomers with the correct ways to avoid all these false hypotheses in future, and all the logical circles which can be committed in this context.[18] But astronomers, by their false hypotheses, usually err only more or less with respect to the motion of planets, whereas physicists may err in the very nature of the thing they discuss. There is nothing simpler than to adjust a particular cause to an effect; and you know that this is familiar to astronomers who, by means of different hypotheses – using either circles or epicycles – come to the same conclusion, and you are familiar with the same thing in your geometry. But in order to prove that the cause of a particular effect is its true and unique cause, one has to prove at least that such an effect cannot be produced by any other cause.

Now I believe that, given who you are, you would not have failed to anticipate all the objections that may be raised against you, in accordance with page 69 of your *Method*,[19] but by still keeping to yourself the detailed knowledge of your principles of physics, from which everything else is deduced, you wished to have fun not only by making keen minds wish for the publication of your physics, but by also providing them with practice in the difficulties that you have left in your new doctrine. Indeed, you even encourage them on page 75 of your *Method*[20] so much that you beseech them to send their objections to you, and that was my main reason for sending you this letter.

Descartes to Father Vatier, 22 February 1638[21]
(AT I, 558–61, 562–5)

I was delighted by your kindness in looking so carefully over the book of my essays, and in giving me your reactions to it with so many signs of your goodwill. When I was sending it to you, I would have enclosed a letter and would have used that opportunity to assure you of my very respectful service, except that I had hoped to circulate the book without making known the name of its author. Since this plan did not succeed, I ought to believe that it was more your affection for the father, rather than the merits of the child, which caused the favourable welcome that you gave it, and I am very much obliged to thank you for that. I am not sure if I am flattering myself, because of many things which are extremely favourable to me in the two letters that I received from you; but I will say frankly that, among all those who were kind enough to tell me what they thought about my writings, none of them, it seems to me, has done me as much justice as you, and none has been as favourable, as impartial and as well-informed about the subject. I am surprised that your two letters were able to follow each other so quickly; and when I saw the first one, I was convinced that I should not expect the second one until after your St Luke holiday.[22]

But in order to give you a detailed reply, I shall say first that my plan was not to teach the whole of my method in the *Discourse* in which I propose it, but only to say enough about it to show people that the novel views that would appear in the *Dioptrics* and the *Meteors* were not thought up casually and that they might deserve to be examined. Also, I was not able to illustrate the use of this method in the three treatises that I have provided, because it prescribes an order for discovering things that is rather different from what I thought should be used in order to explain them. However, I have provided a sample of the method in describing the rainbow and, if you take the trouble to reread it, I hope it will satisfy you more than would have been possible on a first reading, because the subject matter is rather difficult in itself. Now, what made me attach these three treatises to the *Discourse* which precedes them is that I was convinced that they might be enough to convince those who had examined them carefully, and had compared them with what has been written about the same matters previously, that I am using a method which is distinct from what is ordinarily used and that, perhaps, it is not among the worst ones.

It is true that I was very obscure in what I wrote about the existence of God in this treatise on method, and, although that is the most important section, I concede that it is the least developed of the whole work. That results partly from the fact that I decided to include it only at the last minute, when the publisher was rushing me. But the main cause of its obscurity results from the fact that I did not dare discuss at length the arguments of the sceptics, nor say everything required *in order to lead the mind away from the senses* [in Latin]. For it is not possible to know the certitude and clarity of the arguments which prove God's existence, in my way of proving it, except by reminding oneself distinctly of those which make us notice the uncertainty of all our knowledge of material things. It did not seem appropriate to me to put these thoughts in a book in which, I had hoped, even women would be capable of understanding some things,[23] while the most subtle minds would also find enough material to occupy their attention. I also concede that, as you have very well remarked, this obscurity derives partly from the fact that I have assumed that certain notions, which became familiar and clear to me from the habit of thinking, should be the same for everyone; for example, that our ideas can receive their forms or their being only from some external objects or from ourselves, and therefore they cannot represent any reality or perfection which is not either in those objects or in us, and other similar notions. I have considered providing some clarification of this in a second edition.

I certainly thought that what I said I had put in my treatise on light,[24] concerning the creation of the universe, would be incredible. For it is only ten years since I myself would have been unwilling to believe, if someone else had written it, that the human mind would be capable of achieving such knowledge. But my conscience, and the power of the truth, prevented me from fearing to propose something that I believed I could not omit without betraying myself, and of which I already have many witnesses here. Besides, if that part of my physics – which is completed and of which a final copy has been prepared some time ago – ever sees the light of day, I hope that our descendants will not be able to doubt it . . .

As regards light, if you look at page 3 of the *Dioptrics* you will see that I explicitly said there that I would speak only hypothetically about it; in fact, because the treatise which contains the whole body of my physics is entitled *On Light*, and since light is the thing that I explain there more fully and in greater detail than anything else, I did not wish to repeat the same things elsewhere but merely to give some idea of them by means of analogies or hints, in so far as it seemed to me necessary for the subject matter of dioptrics.

I am grateful to you for saying that you were happy that I did not allow myself to be anticipated by others in the publication of my ideas. But that is something I have never feared. For apart from the fact that it is completely irrelevant whether I am the first or the last to write the things that I write, once they are true, all my thoughts are so interconnected and depend so much on each other that one could not appropriate one of them without knowing all of them. Please do not defer telling me about the difficulties you find in what I have written about refraction or about anything else. If you were to wait until the publication of my more detailed views about light, you may be waiting a long time. As regards what I assumed at the beginning of the *Meteors*, I could not demonstrate it a priori except by providing the whole of my physics; but the experiences that I have deduced necessarily from it, and that could not be deduced in the same way from any other principles, seem to me to demonstrate it sufficiently a posteriori. I had certainly anticipated that this way of writing would shock many readers initially, and I thought that I could provide an easy remedy by merely removing the term 'assumptions' from the first things of which I speak, and by acknowledging them only as I provide some reasons to prove them. But I would tell you frankly that I chose this way of proposing my thoughts both because I believed I could deduce them in an orderly way from the first principles of my metaphysics and therefore wished to omit all other kinds of proof, and because I wished to test if the mere exposition of the truth would be enough to make it convincing, without including any disputes or refutations of contrary opinions. Those of my friends who have read my treatises on dioptrics and meteors most carefully assure me that I have succeeded in this; for although they may have found initially just as many difficulties in them as others have, still, once they had read them and reread them three or four times, they said they found nothing which, in their view, could be called into doubt. Indeed, it is not always necessary to have a priori reasons to convince people of the truth; and Thales, or whoever it may have been who said that the moon received its light from the sun, provided no other proof of it except that, by making this assumption, the various phases of the moon can be explained. That has been enough, since that time, for that view to have spread throughout the world without contradiction. Thus my thoughts are related in such a way that I dare to hope that my principles will be found to be proved – as well as the waxings and wanings of the moon prove that it borrows its light – by the consequences which are deduced from them, if people reflect on them enough to become familiar with them, and if they are considered all together.

The only other thing about which I must reply to you concerns the publication of my physics and metaphysics. I can tell you about that in one word. I desire it as much as or more than anyone else, but only in circumstances without which I would be imprudent to want it. I would also tell you that I have no fear, ultimately, that anything contrary to the faith will be found there. On the contrary, I dare to pride myself that the faith has never been as well supported by human reasons as it could be if my principles were adopted. In particular, transubstantiation, which Calvinists claim is impossible to explain by means of the standard philosophy, is very intelligible according to my philosophy.[25] But I see no indication that the circumstances that could oblige me to do so are realized, at least not for a long time; and being content, for my part, to do whatever seems to be my duty, I defer to the providence which rules the world. Since I know that it is that providence which gave me the small beginnings of which you have seen some samples, I hope that it will give me the grace to complete it, if it is useful for its glory; and if it is not, I wish to abstain from desiring it. Finally, I assure you that the sweetest fruit that I have harvested thus far from what I have published is the approval that you were kind enough to provide by your letter. It is particularly pleasant and cherished because it comes from someone of your reputation and your order, and from the very place where I had the good fortune to receive all the teaching of my youth, and which is the home of my teachers, towards whom I remain forever grateful.[26]

Descartes to Reneri [for Pollot], April or May 1638[27]
(AT II, 34–43)

First, it is true that if I had said absolutely that one must maintain whatever opinions one had once decided to adopt, even if they are doubtful, I would be just as reprehensible as if I had said that one should be opinionated and stubborn, because maintaining an opinion is the same as persevering in the judgement that one has made about it. But I said something completely different, namely, that one must be resolute in one's actions even while remaining uncertain in one's judgements (see page 24, line 8),[28] and that one should follow the most doubtful opinions just as decisively – that is, not act with any less resolution on opinions that one judges to be doubtful, once the decision has been made to act on them, i.e. once one has decided that there are no other opinions which one judges are better or more certain – as if one knew they were the best; which indeed is what this view implies (see page 26, line 15).[29] There is no fear that this steadfastness in action would involve us more and more in error and vice, because error can occur only in the understanding which, despite this, I assume is still free and classifies what is doubtful as doubtful. I also apply this rule principally to decisions about living which cannot be deferred, and I use it only provisionally (page 24, line 10);[30] for I plan to change my views as soon as I can find better ones, and I will not pass up any opportunity to search for them (page 29, line 8).[31] Finally, I had to speak about this resolution and firmness concerning actions, both because it is necessary for peace of mind and to prevent people from blaming me for having written that, in order to avoid rashness, once in one's life one must set aside all the opinions that had previously been believed. For apparently it has been objected that such a universal doubt could give rise to great indecision and a great loss of moral control. Thus it seemed to me that I could not have used any more care than I did by placing decisiveness, in so far as it is a virtue, between the two vices which are its contraries, namely indecisiveness and obstinacy.[32]

2. It seems to me that it is not a fiction, but a truth that should not be denied by anyone, that there is nothing which is *completely* in our power apart from our own thoughts, at least if one understands the word 'thought' (as I do) to mean all the operations of the soul; thus, not only meditations and acts of the will, but also the operations of seeing, hearing, deciding to perform one movement rather than another, etc., are all thoughts in so far as they depend on the soul. And there is nothing at all, apart from what is included under this word, which is attributed strictly to human beings in the language of philosophy. For as regards the operations that are attributed to the body alone, they are said to take place in us rather than to be performed by us. By the word 'completely' and what follows from it – viz. when we have done our best in respect of some external things, everything that we fail to achieve is *absolutely* impossible – I indicate clearly enough that I did not mean to say that external things were therefore not at all in

our power, but merely that they are such only to the extent that they can follow our thoughts, and that they cannot do this *absolutely* or *completely* because there are many forces outside us which can hinder the effects of our plans. To express myself more clearly I even joined together these two terms, 'with respect to us' and 'absolutely'; critics could interpret them as if they were mutually incompatible, if they were unaware of the meaning that makes them compatible. Now in spite of the fact that nothing external to us is in our power except in so far as it depends on the command of our mind, and that there is nothing absolutely in our power except our thoughts – and there is no one, its seems to me, who would have any difficulty in accepting that, once they think about it explicitly – I said none the less that one needs to get used to believing it, and that one even needs a lengthy training and a frequently repeated meditation to believe it. The reason is that our appetites and passions constantly tell us the opposite, and we have found so frequently since our infancy that, by crying or demanding, etc., we got our nurses to obey us and we got what we wanted, that we have gradually become convinced that the world was made only for us and that we have a right to everything. Those who were born lucky, into a noble family, have more opportunities to deceive themselves in this way. One also sees that they are usually the ones who are most impatient in coping with misfortunes. But it seems to me that there is no more appropriate role for philosophers than to train themselves to believe what true reason tells them, and to beware of the false opinions of which their natural appetites would convince them.

3. When someone says: 'I am breathing, therefore I am,' and if they try to deduce their existence from the fact that respiration cannot take place without it, they do not deduce anything, because it would be necessary first to have proved that it is true that they are breathing, and that is impossible unless they have also proved that they exist. But if they wish to deduce their exisence from the feeling or belief they have that they are breathing – in the sense that, even if this belief were not true, one still judges that it is impossible to have that view without existing – then they reason very well because this thought of breathing is present to the mind prior to the thought of our existing, and we cannot doubt that we have it as long as we have it (see page 36, line

22).³³ Thus when someone says: 'I am breathing, therefore I am,' this is equivalent, in that sense, to: 'I am thinking, therefore I am.' And if one thinks about it, one finds that all the other propositions from which we can deduce our existence in that way amount to the same thing. Thus, by using them, one does not prove the existence of the body (that is, the existence of a nature which occupies space, etc.) but only that of the soul (that is, of a nature which thinks). Although one could wonder whether it is not the same nature that thinks and occupies some space, that is, one which is both intellectual and corporeal together, nevertheless one knows it only as intellectual by means of the path that I have proposed.

4. From the fact alone that we conceive clearly and distinctly of the two natures of the soul and the body as different, it follows that they are truly different, and therefore that the soul, without the body, is capable of thinking despite the fact that, as long as it is united with it, its operations can be disturbed by the poor condition of the body's organs.

5. Even though the pyrrhonists³⁴ have deduced nothing certain from their doubts, that does not imply that it cannot be done. And I would try to show here how one could use these doubts to prove God's existence, by clarifying the difficulties which remain in what I have written about it. But I have been promised that I would soon be sent a summary of everything that can be doubted about this question, and that may give me an opportunity to do so better. That is why I ask whoever made these comments to allow me to wait until I have received the summary.

6. It is certain that the similarity between most of our actions and those of beasts has given us, from the beginning of our lives, such an opportunity for thinking that beasts act as a result of an inner principle which is similar to that which is present in us – that is, by means of a soul, which has sensations and passions like ours – that we all think about this view naturally. Whatever reasons one might have for denying it, they can hardly be expressed publicly without leaving oneself open to the ridicule of children and feeble minds. But for those who wish to know the truth, they should above all distrust views which they have been told about since their infancy. And to find out what should be

believed about it, it seems to me that one should consider what judgement would someone make who had been reared for their whole life in some place where they had seen no other animals apart from human beings and where, being committed to the study of mechanics, they would have constructed or assisted in the construction of many automata, some of which were shaped like humans, others like horses, dogs, birds and so on, and which walked, ate, breathed and, in a word, imitated as much as possible all the other actions of the animals that they resembled, including even the signs we use to express our passions (such as crying when struck, fleeing when a loud noise is made near by, etc.). Thus such a person would find it difficult to distinguish between real human beings and those who only had the same shape; and experience would have taught them that, in order to recognize real human beings, there are only the two signs that I explained on page 57 of my *Method*.[35] The first is that automota never use words or even signs to reply to the questions they are asked, except by chance. The other sign is that, while the movements they make are often more regular and certain than the wisest human beings, they are still lacking, more than the most insane people would be, many things they would have to do in order to imitate us. We have to consider, I claim, what judgement would be made by this person about the animals that are in our environment if they were to see them, especially if they were endowed with knowledge of God or at least had noticed how inferior is all the work that human beings display in their artifacts in comparison with what nature displays in the composition of plants, in so far as nature fills them with an infinity of little conduits which are unobservable to the naked eye, through which it causes certain liquids to rise gradually. Once they reach the top of their branches, they mix together, combine harmoniously and dry out in such a way that they form leaves, flowers and fruits. If such an observer firmly believed that if God or nature had formed some automata which imitated our actions, they would imitate them more perfectly and would be incomparably better constructed than any that could be invented by human beings. Now there is no doubt that this observer, seeing the animals in our environment, and noticing in their actions the two main things that make them different from ours – that they would also have got used to noticing in their automata – would not judge that there is any genuine feeling in them nor any real passions, as in us, but that they are merely automata which, composed by nature, would be incomparably more skilled than any of those that they themselves had made previously. Thus it only remains to consider whether this judgement – which would be formed with knowledge of the subject matter, and without having been prejudiced by any false views – is less credible than the one we have made since our infancy and have retained since then merely as a result of habit. For it is based only on the similarity between some external actions of beasts and our own; and it is not at all adequate to prove that there is a corresponding similarity in our inner actions . . .

11. It is well known that I do not claim to show that water particles have the same shape as certain animals, but merely that they are long, slippery and flexible.[36] For if one could find some other shape by which all their properties could be explained equally well, I would be willing to accept it. But if that cannot be done, I can see no objection to imagining them having this shape as much as any other, since they must have some shape or other, and this one is among the most simple.

Descartes to Mersenne, 27 May 1638
(AT II, 141–4)

You ask whether I believe that what I have written about refraction is a demonstration. I believe it is, at least in so far as it is possible to provide a demonstration in this subject matter without having first demonstrated the principles of physics by means of metaphysics – something that I hope to do some day, but it has not been done yet – and in so far as any other question in mechanics, or optics, or astronomy, or any subject which is not purely geometrical or arithmetical, has even been demonstrated. But to demand geometrical demonstrations from me, in something which depends on physics, is to expect me to do the impossible. If one wishes to apply the term 'demonstration' only to geometrical proofs, then one must say that Archimedes never demonstrated anything in mechanics, nor Witelo[37] in optics, nor Ptolemy in astronomy, and so on – but that is not what is said. For in these disciplines one is satisfied if the authors presuppose certain things which are not manifestly incompatible with experience and then speak consistently, without committing any logical mistakes, even if their initial assumptions were not exactly true. For example, I could demonstrate that even the definition of the centre of gravity that was given by Archimedes is false, and that there is no such centre, and that the other things he assumes elsewhere are not exactly true either. As regards Ptolemy and Witelo, they have assumptions that are much less certain, but none the less one should not for that reason reject the demonstrations that they deduce from them. Now what I claim to have demonstrated about refraction does not depend on the truth about the nature of light, nor on the fact that light does or does not appear in an instant, but only on my assumption that it is an action or power which follows the same laws as local motion, in the way in which it is transmitted from one place to another, and that it is transmitted by means of a very subtle fluid which is present in the pores of transparent bodies. As regards your objection to the claim that it is transmitted in an instant, the term 'instant' is equivocal. You seem to think of it as denying every kind of priority, so that the light of the sun could be produced here on earth without first passing through the whole space between us and the sun; whereas the term 'instant' excludes only temporal priority, and does not prevent all the lower parts of the ray of light from being dependent on the higher parts, in the same way as the end of a successive motion depends on all its preceding parts. You should know that there are only two ways to refute what I have written: one of them is to prove by means of some experiences or reasons that the things that I have assumed are false; the other is to show that what I deduce from them cannot be deduced in that way. Mr Fermat[38] understood this very well; for that is how he tried to refute what I have written about refraction, by trying to show that it contained a logical mistake. But for those who simply say that they do not believe what I have written because I deduce it from certain assumptions that I have not proved, they do not know what they are asking for, nor what they ought to ask for.

Descartes to Morin, 13 July 1638
(AT II, 197–200)

The objections that you have taken the trouble to send me are such that I would have welcomed them from anyone; but your esteemed position among the learned, and the reputation which your writings have earned you, make them much more acceptable to me when they come from you rather than from someone else. The best sign of this that I can give, I think, is the care I take here in replying to them in detail.

You begin with my assumptions, and you say that 'the apparent movements of the heavens are deducible just as certainly from the hypothesis that the earth is stationary as from that of its motion', which is something that I readily accept. I had wished that what I wrote in the *Dioptrics* about the nature of light would be understood in the same way, so that the strength of the mathematical demonstrations, which I tried to provide there, would not depend on any opinion in physics, as I made sufficiently clear on page 3.[39] If light can be imagined in some other way, so that all the properties that are known by experience are explained, one will see that everything I have demonstrated about refractions, vision and the rest could be deduced from it just as much as from the hypothesis that I proposed.

You also say that 'proving effects by a cause, and then proving this cause by the same effects is a vicious circle',[40] which I accept. But I do not accept, for that reason, that it would be a vicious circle to explain effects by a cause and then to prove the cause by the effects, because there is a big difference between proving and explaining. I add that one can use the term 'demonstrate' to mean one or the other, at least if it is used according to common usage and not with the special meaning that the philosophers give it. I also add that it is not circular to prove a cause by many effects which are known independently and then, reciprocally, to prove certain other effects by means of this cause. I included these two meanings together on page 76 in the following words: 'As the last ones are demonstrated by the first which are their causes, the first arguments are demonstrated reciprocally by the last which are their effects.'[41] But I should not be accused of speaking ambiguously because of that, for I explained myself immediately afterwards by saying that 'since experience makes most of these effects very certain, the causes from which I deduce them are used not so much to prove as to explain them; but, in exactly the opposite way, it is the former which are proved by the latter'. And I put 'they are used not so much to prove them' instead of 'they are not used at all', so that people would know that each of these effects could also be proved by this cause if they happened to be called into doubt, on condition that the cause had been proved by other effects. I do not see that I could have used other terms than I did here in order to explain myself better.

You also say that 'astronomers often make assumptions which cause them to fall into great mistakes, such as, for example, when they assume mistakenly the parallax, the obliquity of the ecliptic, etc.'. I reply to this that those things are never included among the kind of assumptions or hypotheses that I spoke about, and that I clearly identified the kind in question by saying that 'one could draw very true and very certain conclusions from them even though they are false or uncertain'. But the parallax or the obliquity of the ecliptic cannot be assumed to be false or uncertain, but only true; whereas the equator, the zodiac, epicycles and other similar circles are usually assumed to be false, and the mobility of the earth is assumed to be uncertain, but that does not prevent us from deducing very true things from them.

Finally, you say that 'there is nothing simpler than to adjust a particular cause to an effect'. But although there may indeed be many effects to which it is easy to adjust different causes, with a different cause for each effect, it is not so easy to adjust one and the same cause to many different effects if it is not the real cause from which they result. There are often even some effects such that, by proposing one cause from which they can be deduced clearly, this is enough to prove that it is their true cause. I claim that all those I spoke about are of this kind. For if one considers that, in everything that has been done to date in physics, people have only tried to imagine some cause by which natural phenomena could be explained, even though they were hardly able to succeed; and if one then compares other people's hypotheses with mine, that is, all their real qualities, substantial forms, elements and things like that, which are almost infinite in number,[42] with the single hypothesis that all bodies are composed of some parts, which is something that can be seen with the naked eye in many cases and can be proved by innumerable reasons in the case of others (for as regards my other assumption, viz. that the parts of one body or another have one shape rather than another, it is easy to demonstrate that to those who agree that bodies are composed of parts); and finally, if one compares what I have deduced from my assumptions concerning vision, salt, the winds, the clouds, snow, thunder, the rainbow and similar things with what others have deduced from their assumptions about the same phenonema, I hope that will be enough to convince those who are not too prejudiced that the effects that I explain have no other causes apart from those from which I deduced them, even if I defer providing a demonstration to some other place.

Descartes to Debeaune, 30 April 1639[43]
(AT II, 542–4)

First, I hold that there is a definite quantity of motion in the whole of created matter, and that it never increases or decreases. Thus, when one body makes another body move, it loses as much of its own motion as it gives to the other one; for example, when a stone falls to earth from a height, if it stops and does not bounce back again, I conceive of that happening because it shakes the earth and thus transfers its motion to the earth. But if the portion of the earth that it moves contains one thousand times more matter than itself, then by transferring all its motion it gives the earth only a one-thousanth part of its speed. Therefore if two unequal bodies receive the same amount of motion each, and if this equal quantity of motion does not impart as much speed to the larger one as it does to the smaller of the two, one can say that the more matter a body contains the more 'natural inertia' it has in this sense of the term. One could add that, when a body is large, it can transfer its motion to other bodies more easily than a small body can, and that it can be moved by them less easily. Thus there is a type of 'inertia' which depends on the quantity of matter, and another type which depends on the surface area of bodies.

As regards weight, I do not imagine it is anything other than all the subtle matter between us and the moon rotating rapidly around the earth and pushing towards the earth all the bodies which cannot move as quickly. Now when they have not begun to descend, it pushes them with more force than when they have already begun to do so. For if it happens that they descend as quickly as it moves them, it will not push them at all, and if they descend more quickly, it will resist them. You can see from this that there are many things to consider before being able to decide anything about speed, and that is what has always put me off doing so. But one can also explain many things by means of these principles which could not have been explained previously. Moreover, I would not write to you as freely about these things – which I was unwilling to discuss elsewhere, because proving them presupposes my *World* – if I had not hoped that you would understand them favourably, ...

Descartes to Mersenne, 27 August 1639
(AT II, 570–71)

I finally received two copies of the book entitled *On Truth*, which you were kind enough to send me.[44] I shall give one copy to Mr Bannius[45] on your behalf as soon as possible, because I assume that was what you intended. I have no time to read it now; for that reason I cannot tell you anything about it except that, when I read it earlier in Latin, I found many things at the beginning that seemed very good to me, and the author shows that he is better informed than usual about metaphysics (which is a science that hardly anyone understands). But it seemed to me that he later mixed together religion and philosophy, and that is completely contrary to my understanding. I have not finished the book, which is something I hope to do as soon as I have the time to read a book, and I will also look at the *Philolaus*.[46] Meantime, however, I am studying without using a book.

Descartes to Mersenne, 16 October 1639
(AT II, 596–9)

Since my last letters, I took time to read the book that you were kind enough to send me;[47] and since you asked me what I thought of it and it deals with a subject on which I have worked all my life, I think I should write about it to you in this letter. I find many good things in it, *but not to everyone's taste* [in Latin]; for there are few people who are capable of understanding metaphysics, and in the body of the book the author follows a line of argument that is very different from the one that I have followed. He examines what truth is; for my part, I have never had any doubts about it, since it seemed to me to be a notion that is so transcendentally clear that it is impossible not to know it. In fact, while there are many ways of examining a balance before using it, there would be no way to learn what truth is if it were not known naturally. For what reasons would we have for agreeing to anything that could teach us what truth is, if we did not know that the thing in question was true, that is, if we were not acquainted with truth? Thus one can easily explain *what a word means* [in Latin] to those who do not know a language and one can tell them that this word 'truth', when used in a strict sense, denotes the conformity of thought with the object, but that when truth is attributed to things which are outside our thought, it means simply that these objects can serve as objects of true thoughts, either of our own thoughts or those of God. But one cannot provide any logical definition which helps us to know its nature. I believe the same about many other things, which are very simple and are known naturally, such as shape, size, motion, place, time, etc., so that, when one tries to define these things, one obscures them and confuses oneself. For example, someone who walks across a room gives a much better explanation of what motion is than someone who says that it is *'the act of a being in potency, in so far as it is in potency'* [in Latin],[48] and likewise for the others.

The author takes universal agreement as a criterion of his truths; for my part, I have only the natural light as a criterion for mine, which has something in common with his; for since everyone has the same natural light, they should all apparently have the same notions. But my criterion is very different, because there is hardly anyone who uses this natural light well and thus many people may agree in accepting the same error (for example, all our acquaintances), and there are many things that can be known by the same natural light but no one has so far reflected on them.

He claims that there are as many faculties in us as there are different things to know. I can understand this only in the same way as if one said that, because wax can receive an indefinite number of shapes, it has an indefinite number of faculties to receive them. In that particular sense, his claim is true; but I see no advantage in speaking that way, and it seems to me rather that it could be harmful by providing the ignorant with an opportunity for imagining as many different small entities in our soul. For that reason, I would prefer to think that the wax, simply as a result of its flexibility, receives all kinds of shapes, and that the soul acquires all its knowledge as a result of the reflection it performs, either on itself in the case of intellectual things, or on the various dispositions of the brain to which it is joined in the case of physical things, whether these dispositions depend on the senses or on other causes. But it is very useful not to accept anything into one's beliefs without considering for what reason or cause one accepts it, which comes back to what he says: that one should always consider which faculty one is using, etc.

There is also no doubt that, as he says, to avoid being deceived by the senses one must take care that there is nothing lacking on the part of the object, the medium, or the organism, etc.

The World,
or a Treatise on Light and the Other
Principal Objects of the Senses

(1633; published 1664)

NOTE ON THE TEXT

Descartes had originally written this version of his physics in two parts, in French, as a treatise on light and a treatise on man. It was ready for publication in 1633, when news of Galileo's condemnation by the Roman Inquisition reached the Netherlands. Since Descartes was proposing the heliocentric system of astronomy which provoked the Inquisition's judgement, he decided to shelve his book and to turn his attention instead to other questions that were less likely to bring him into conflict with the teaching authority of the church to which he belonged. The text was published posthumously in two parts, and appeared in reverse order as *De Homine* (Leiden, 1662) and *Le Monde de Mr Descartes, ou le Traité de la Lumière et des autres principaux objets des Sens* (Paris, 1664). The first of these was a Latin translation by F. Schuyl of Descartes's treatise on man; an edition of the French text of the same treatise, with extensive notes by Louis de la Forge, was published in Paris in 1664. The text included here contains the first seven chapters of *Le Monde*; they are translated from vol. XI of the Adam and Tannery edition of Descartes's works.

THE WORLD

CHAPTER I

The Difference between Our Sensations and the Things That Produce Them

Since my plan here is to discuss light, the first thing that I want to bring to your attention is that there may be a difference between our sensation of light, i.e. the idea which is formed in our imagination by means of our eyes, and whatever it is in the objects that produces that sensation in us, i.e. what is called 'light' in a flame or in the sun. For although everyone is commonly convinced that the ideas we have in our thought are completely similar to the objects from which they originate, I see no argument that guarantees that this is so; on the contrary, I am aware of many experiences which should make us doubt it.

You are well aware that words do not in any way resemble the things they signify; that does not prevent them from causing us to think about those things, often without us even noticing the sound of the words or their syllables. Thus it can happen that we hear a speech and understand its meaning very well, but we are unable later to say what language was used in making the speech. Now if words – which have meaning only as a result of a human convention – are enough to make us think about things that do not resemble them in any way, why is it not possible that nature may also have established a particular sign which would make us have the sensation of light, even though such a sign contains nothing in itself that resembles the sensation? And is this not the way in which it has established laughter and tears, to make us read joy and sorrow on people's faces?[1]

But you will object, perhaps, that our ears cause us actually to sense only the sound of words and that our eyes give us a sensation only of whoever laughs or cries, and that it is our mind – since it has remembered what those words or what such a countenance signify – which simultaneously represents their meaning. I could reply that, nevertheless, it is our mind that represents the idea of light to us every time the action which signifies it impinges on our eye. But without wasting time in disputes, I would be better off introducing a different example.

Even when we are not paying attention to the meaning of words and we hear only the sound of them, do you think that the idea of this sound which is formed in our thought is something that resembles the object that causes it? Someone opens their mouth, moves their tongue and breathes out; I see nothing in all these actions which is not very different from the idea of sound that they cause us to imagine. Most philosophers maintain that sound is merely a particular vibration of the air which comes to strike our ears; consequently, if our sense of hearing reported to our thought the real image of its object, then instead of causing us to think of sound it would have to make us think of the motion of the particles of air which are vibrating at that time against our ears. However, since everyone may not wish to believe what philosophers say, I shall introduce a different example again.

Touch is, among all our senses, the one that is thought to be least deceptive and most reliable. Thus if I show you that even touch makes us conceive many ideas that do not in any way resemble the objects which produce them, I do not think you should find it strange if I say that the sense of sight may do the same. Now there is no one who does not know that the ideas of tickling and of pain, which are formed in our thought when external bodies touch us, bear no resemblance at all to such bodies. One rubs a feather lightly over the lips of a sleeping child and they have a sensation of being tickled; do you think that the idea of tickling which they conceive resembles something in the feather? A soldier returns from battle; during the heat of the fray he may have been wounded without being aware of it. But now when he begins to cool off he feels pain, and believes he is wounded. A surgeon is summoned, the soldier's armour is removed, he is examined and it is discovered eventually that what he felt was nothing more than a buckle or strap which was caught under his armour and pressed against him, thereby causing discomfort. If his sense of touch, when it was causing him to feel the strap, had impressed an idea of the strap on his thought, he would not have needed a surgeon to tell him what he felt.

Now I see no reason to make us believe that, whatever is in the objects from which we get a sensation of light, resembles this sensation any more than the actions of a feather or a strap resemble tickling or pain. At the same time, I have not introduced these examples to make you believe unconditionally that light in objects is something different from what it is in our eyes; my objective was only to make you doubt it so that, guarding against being convinced of the opposite view, you could better investigate with me what light is.

CHAPTER 2

What Do the Heat and Light of Fire Consist In?

I know only two kinds of bodies in the world in which light occurs, namely, the stars, and a flame or fire.[2] Since the stars are undoubtedly further removed from human knowledge than fire or flame, I shall try to explain first what I notice about flames.

When a flame burns wood or some other similar material, we can see with the naked eye that it moves the small parts of the wood and separates them from one another, thus transforming the finer parts into fire, air and smoke and leaving the larger parts behind as ashes. Thus someone else may imagine, if they wish, the 'form' of fire, the 'quality' of heat, and the 'action' which burns it, as things that are completely distinct in the wood.[3] For my part – as someone who is afraid of making a mistake if I assume anything more in the wood than what I see must necessarily be there – I am satisfied to conceive in it the movement of its parts. You may posit 'fire' and 'heat' in it, and make it burn as much as you wish; but if you do not also assume that some of its parts move and become detached from those which are next to them, I could not imagine that the wood could undergo any change or alteration. On the other hand, take away the 'fire' and the 'heat' and prevent it from burning; on condition simply that you grant me that there is some power which violently moves its finer parts and separates them from the larger parts, I find that this alone could cause it to undergo all the same changes which are observed when it burns.

Now in so far as it seems to me impossible to conceive of a body being able to move another except by also moving itself, I draw the conclusion that the body of the flame which acts on the wood is composed of small parts which move independently of each other, with a motion which is very rapid and violent, and that, by moving themselves in this way, they push and move with themselves those parts of the wood that they touch and that do not offer them too much resistance. I claim that the parts of the flame move independently of each other; for although it often happens that many of them act together in harmony to produce the same effect, we see nevertheless that each of them acts on its own on the bodies that they touch. I claim that their motion is very rapid and violent; for, since they are so small that it would be impossible to distinguish them by sight, they would not have as much force as they have to act on other bodies if the speed of their motion did not compensate for what they lack in size.[4]

I am not adding anything about the direction in which each part of the flame moves. For if you consider that the power [of a particle] to move itself and the power which determines the direction in which it must move are two completely distinct things, and that one may exist without the other (as I explained in the *Dioptrics*[5]), you will easily judge that each part moves in the manner that is made least difficult for it by the arrangement of the surrounding bodies, and that in the same flame it is possible to have some parts which go upwards and others which go downwards, some in a straight line, in a circle, and in every direction, without thereby changing anything in its nature. Thus if you see almost all of them tending upwards, it is not necessary to think that there is any reason for this, except that the other bodies which touch them are almost always disposed to resist them more in all the other directions.

But once it is acknowledged that the parts of the flame move in this way, and that it is enough to conceive of their movements to understand how the flame has a power to consume wood and to burn, please let us examine if the same thing would not also be enough to make us understand how it heats us and how it illuminates. For if that is found to be the case, it will not be necessary for the flame to have any other 'quality' and we can say that it is this motion alone that is sometimes called 'heat' and sometimes 'light', depending on the different effects that it produces.

As regards heat, it seems to me that the sensation we have of it can be taken as a kind of pain when it is violent, and that it can be taken sometimes as a kind of tickling when it is moderate. Since we have already said that there is nothing outside our thought that is similar to the ideas we have of pain and tickling, we may well believe also that there is nothing which is similar to our idea of heat. Rather, everything which can in various ways move the small parts of our hands or of any other part of our body is able to stimulate this sensation in us. Many experiences also support this view, because our hands are warmed simply by rubbing them together; and any other body can be warmed without being put near a fire, on condition that it is shaken and moved in such a way that some of its small parts move and they can thereby move the small parts of our hands.

As regards light, one can easily conceive likewise that the same motion which is in the flame is enough to make us have a sensation of light. But since that is the main part of my project, I want to try to explain it at length when I return to this topic later.

CHAPTER 3
Hardness and Liquidity

I believe that there is an infinity of different motions in the world which last for ever. After having noticed the greatest among them – which cause the days, months and years – I take note that the vapours of the earth never cease to ascend to the clouds and descend again, that the air is constantly moved by winds, that the sea is never at rest, that fountains and rivers flow incessantly, that the most stable buildings eventually fall into decay, that animals and plants are either growing or decaying – in brief, that there is nothing anywhere which does not change. From this I know clearly that it is not only in a flame that there are many small parts which do not cease to move, but that the same motion occurs in all other bodies, even if their actions are not as forceful and even if, because of their small size, they cannot be perceived by any of our senses.

I shall not stop to look for the cause of their motions; for it is enough for me to think that they began to move as soon as the world began to exist and, if that is accepted, I find by reasoning that it is impossible for their motions ever to cease or even to change in any way, apart from changing the subject in which they are present. That is to say, it may well be the case that the force or power to move itself, which is found in one body, passes wholly or partly into another body and thus is no longer in the first body; but it cannot cease to exist entirely in the world. I am sufficiently convinced, I claim, by my arguments about this point, but this is not the appropriate occasion to tell you about them. Meantime, you may imagine if you wish – as most of the learned do – that there is some prime mover which, by revolving around the world at an inconceivable speed, is the origin and source of all the other movements that occur there.

Now, as a result of that consideration, there is a way to explain the cause of all the changes that occur in the world and of all the variations which appear on earth. But I shall be content here to speak only about those that are relevant to my subject.

The first thing I would like you to note is the difference between hard bodies and those which are liquid, and in order to do that, consider that every body can be divided into extremely small parts. I do not wish to decide if their number is infinite or not; however, it is certain that, at least from the point of view of our knowledge, their number is indefinite, and we may assume that there are many millions of them in the smallest grain of sand that can be perceived by the naked eye.

Note that if two of these small parts are in contact with each other, while not in the process of moving apart, some force is needed, no matter how small, in order to separate them. For once they are thus positioned, they would never be inclined of their own accord to arrange themselves differently. Notice also that twice as much force is required to separate two of them than is required for one, and a thousand times more is required in order to separate a thousand of them. Thus if it is necessary to separate many millions of them, all at the same time, as is required perhaps to break a single hair, it is not surprising if a reasonably perceptible force is required.

On the other hand, if two or more of these small parts merely touch one other in passing, while they are in the process of moving in different directions, it is certain that less force would be required to separate them than if they completely lacked all movement, and that even no force at all would be required if the motion with which they are capable of separating on their own is equal to or greater than that with which one wishes to separate them. Now I find no other difference between hard bodies and liquid bodies, except that the parts of the former can be separated from each other much more easily than those of the latter. Thus to compose the hardest body one could imagine, I think it would be enough if all its parts touched each other without leaving any space between them and without any of them being in the process of moving. For what glue or cement could one imagine apart from that, in order to make them hold on to each other more?

I also think that, to compose the most liquid body one could find, it is enough if all its smallest parts move away from each other in the most diverse ways and as quickly as possible, although they are also still able to touch each other on all sides, and to arrange themselves in as small a space as if they were stationary.[6] Finally, I believe that every body is more or less close to these extremes, depending on whether its parts are more or less in the process of separating from each other. And all the experiences that I observe confirm me in this view.

A flame, about which I already said that all its parts are in perpetual motion, is not only liquid but it also makes most other bodies liquid. Notice also that, when it melts metals, it acts with only the same power with which it burns wood. But because the parts of metals are more or less all equal in size, a flame cannot move one of them without moving others, and thus it turns metals into completely liquid bodies. However, the parts of wood are so unequal in size that a flame can separate out the smallest ones and make them liquid – i.e. it can make them fly off in smoke – without moving the larger parts in the same way.

Next to the flame, there is nothing as fluid as air and one can see with the naked eye that its parts move away from each other. For if you take the trouble to look at the small bodies which are commonly called 'atoms' and which appear in the rays of the sun, you will see them flittering about incessantly hither and thither, even when there is no wind to move them. One can also experience the same thing in all thicker liquids, if one mixes liquids of different colours together so as to distinguish their movements better. Finally, this can be observed very clearly in acids, when they move and separate the parts of some metal.

But you could ask me, at this stage: if the motion of the parts of a flame alone is enough to make it burn and make it liquid, why is it that the motion of the parts of air (which makes it also extremely liquid) gives it no power nevertheless to burn but, on the contrary, makes it such that our hands can hardly sense it? I answer that one needs to consider not only the speed but also the size of the parts that are in motion, and that it is the smallest parts which make bodies most liquid, although it is the larger parts which have more force to burn and, in general, to act on other bodies.

Notice in passing that, in this context, I understand a single *part* (and shall always do so henceforth) as everything that is found united together and is not in the process of separating, although even those which are very small can easily be divided into many other smaller parts. Thus a grain of sand, a stone, a rock, and even the whole earth, may henceforth be taken as a single part, as long as we consider in them only one simple and completely equal motion.

Now among the parts of the air, if there are some which are very large in comparison with others, such as the 'atoms' that are visible in it, they also move very slowly; and if there are some which move more quickly, they are also smaller. But in the case of a flame, if some of its parts are smaller than those of air, there are also some which are larger or, at least, there is a larger number of them which have the same size as the largest parts of air and also move more quickly. It is these latter parts alone which have a power to burn things.

CHAPTER 4

That flames contain some smaller parts can be hypothesized from the fact that they penetrate many bodies whose pores are so narrow that even air cannot enter them. That they contain either larger parts, or equally large parts in greater numbers, can be seen clearly from the fact that air alone is not enough to keep a flame burning. That the flame's parts move more quickly is adequately experienced in the forcefulness of their motion. Finally, that it is the largest of these parts which have the power to burn, and not the others, is apparent from the fact that the flame which emerges from brandy or from other very subtle bodies hardly burns at all, whereas, on the contrary, the flame which originates in hard, heavy bodies is very hot.

CHAPTER 4

The Vacuum, and How It Happens That Our Senses Do Not Perceive Certain Bodies

But it is necessary to examine in greater detail why air, despite the fact that it is a body just as much as other bodies are, cannot be sensed as well as they can. In this way we can free ourselves from a mistake with which we have all been prejudiced since our infancy, when we believed that there were no bodies surrounding us apart from those that could be sensed, and therefore, if the air were a body – since we perceive it very slightly – at least it ought not to be as material or as solid as those that we sense more.[7]

On this topic I would like you to notice, first, that all bodies, both hard and liquid, are made from the same matter, and that it is impossible to conceive that the parts of this matter might ever compose a more solid body or one which occupies less space than they do when each one is touched on all sides by the others that surround it. It follows, it seems to me, that if it were possible to have a vacuum somewhere, it ought to be found in hard bodies rather than in those that are liquid.

For it is evident that the parts of the latter can press and act on one another much more easily, because they are in motion, than the stationary parts of other bodies.

If you are putting powder in a jar, for example, you shake it and tap against the jar to make it go in better; but if you are pouring some liquid into the jar, it arranges itself immediately in the smallest space in which it can fit. Even if you consider, in this context, some of the experiences that philosophers have commonly used to show that there is no vacuum in nature, you will easily discover that all the spaces that people think of as empty, and in which we perceive only air, are at least as full, and are full of the same matter, as those in which we perceive other bodies.

Please tell me what would show that nature makes the most heavy bodies ascend and that it breaks open hard bodies – as one sees it doing in certain machines – rather than allow any of its parts to cease to surround one another or to touch other bodies and, at the same time, that it may allow the parts of air (which are so easy to bend and can be arranged in every way) to remain next to one another without being surrounded on all sides or without there being some other body among them which they would touch? Could one really believe that the water in a well should ascend against its natural inclination just so that the pipe in the pump is filled, and could one really think that water in clouds should not descend in order to succeed in filling the empty spaces here below, if there were the slightest vacuum between the parts of the bodies which they contain?

But you could raise a rather significant objection here: that the parts which compose liquid bodies cannot, it seems, move incessantly as I claimed they did if there were no empty spaces between them, at least in the places which they vacate as they move about. I would have some difficulty replying to this if I had not realized, by means of various experiences, that all the movements that occur in the world are in some sense circular. In other words, when a body leaves its place it always enters that of another, and this in turn enters that of a third, and so on until the last one occupies, at the same moment, the place that was vacated by the first body. Thus there is no more of a vacuum between them when they are moving than when they are stationary. Notice here that this does not presuppose that all the parts of bodies, which move together, are arranged precisely in a ring as a genuine circle, or that they all have the same size or shape; for these inequalities can easily be compensated by other inequalities that occur in their speeds.

Now when bodies move in air we do not usually observe these circular movements, because we are used to conceiving of the air only as an empty space. But look at fish swimming in the pool of a fountain; if they do not come too close to the surface of the water, they will not make it move at all even though they pass beneath the surface at great speed. It seems clear from this that the water in front of them which they displace does not push all the water in the pool indiscriminately, but only that which is more conducive to the circles of the fishes' movement and which can return to the place that they vacate. This experience is enough to show how easy and familiar these circular motions are in nature. But I now wish to introduce another experience to show that no motion ever occurs that is not circular. When wine in a cask does not flow out through an opening in the base, because the top is closed completely, it is a misdescription to say – as is usually done – that this happens because of a fear of a vacuum. It is well known that this wine has no mind with which to fear anything; and even if it had, I do not know why it would dread this vacuum, which in reality is merely a chimera. But one should say instead that the wine cannot leave the cask because the outside is as full as it could be, and that the part of the air whose place it would occupy if it descended cannot find another place in the whole universe into which it can go, if one does not make an aperture in the top of the cask through which this air could move in a circle and take its place.

Still, I do not wish to claim that, for this reason, there is absolutely no vacuum in nature. I would be afraid that my discourse might become too long if I undertook to explain the whole issue, and the experiences I spoke about are not enough to prove it, although they are enough to convince us that the spaces where we perceive nothing are filled with the same matter, and contain at least as much of that matter, as those that are occupied by the bodies that we perceive. Thus when a vessel is full of lead or gold, it does not therefore contain more matter than when we think it is empty. This may seem strange to many people whose reason does not reach much further than their fingers, and who think that there is nothing in the world apart from what they touch. But once you have briefly considered what makes us have a sensation of a body, or makes us not sense it, I am certain that you will find nothing incredible in what I have said. For you will realize clearly that, far from the things that surround us being such that they can all be sensed, on the contrary, those that are more usually present can be sensed least, and those that are always present can never be sensed.

The heat of our heart is quite high, but we do not feel it because we are used to it. The weight of our body is not insignificant, but it does not cause us any discomfort. We do not even feel the weight of our clothes because we are used to wearing them. The reason for this is clear enough; for it is certain that we would not be able to perceive any body if it were not the cause of some change in our sensory organs, that is, if it did not in some way move the small parts of the matter of which these organs are composed. Objects that are not always present can easily do this, on condition only that they have enough force; and if they damage something in our sense organs while they are acting on them, that can be repaired subsequently by nature when they are no longer acting. But as regards those things that touch us constantly, if they ever had enough power to produce some change in our senses, and to move some parts of the matter in our senses, with the force to move them they ought to have separated them completely from the other parts at the very beginning of our life; therefore they can only have left behind those that completely resist their impact and by means of which they cannot in any way be sensed. Thus you can see that it is not surprising that there are many spaces around us in which we sense no bodies, even though they do not contain fewer bodies than those in which we sense them the most.

But one should not conclude from this that the gross air which we inhale into our lungs when breathing, which is converted into wind when it is moved, which appears hard to us when enclosed in a balloon, and which is composed only of exhalations and of smoke, is as solid as water or earth. One should follow the common opinion of philosophers, all of whom claim that it is more rare. This can easily be known from experience. When the parts of a drop of water are separated from one another by the motion of heat, they can compose much more of this air than could be contained in the space occupied by the water drop. It follows necessarily that there is a large number of small gaps between the parts of which the water vapour is composed; for there is no other way of conceiving of a rarefied body. But because these gaps cannot be empty, as I have said above, I conclude from all this that there are necessarily other bodies, one or many, mixed in with the air, which fill as tightly as possible the little gaps that the air leaves between its parts. It only remains to consider what these other bodies may be, and after that I hope it will not be difficult to understand what the nature of light may be.

CHAPTER 5

The Number of the Elements, and Their Qualities

Philosophers claim that there is a certain air which is much more subtle than ours above the clouds and that, unlike ours, it is not composed of terrestrial vapours but is a distinct element in its own right. They say that, above the air, there is yet another body which is much more subtle still, which they call the element of fire. They also add that these two elements are mixed with water and earth in the composition of all lower bodies.[8] Thus I would be merely following their opinion if I claim that this more subtle air and this element of fire fill the gaps between the parts of the gross air that we breathe in such a way that these bodies, intermingling with each other, compose a mass which is as solid as any body could be.

But in order to explain my thought on this subject, and so that you may not think that I wish to force you into believing everything that philosophers tell us about the elements, I have to describe them in my own way.

I conceive of the first one, which could be called the element of fire, as the most subtle and penetrating liquid in the world. Following what was said above about the nature of liquid bodies, I imagine that its parts are much smaller and that they move much more quickly than any parts of other bodies. However, lest I be forced to admit any vacuum in nature, I do not attribute parts that have any size or determinate shape to this element, but I am convinced that the impetuosity of its motion is enough to divide it in every way and in every direction by colliding with other bodies, and that its parts change shape at every moment to fit the shape of the places they enter. Thus there is never a gap so narrow, nor a space between the parts of other bodies so small, that the parts of this element do not penetrate them effortlessly and fill them exactly.

As regards the second element – which could be understood as the element of air – I also conceive of it, in comparison with the third element, as a very subtle fluid. But if it is compared with the first element, one has to attribute some size and shape to each of its parts and to think of them all as almost spherical, and as joined together in the same way as grains of sand or dust. Thus they cannot arrange themselves as well nor press against one another in such a way that very small gaps do not always remain between them, and it is much easier for the first element to slide into these gaps than for the parts to change their shape simply in order to fill them. Therefore I am convinced that this second element cannot be so pure in any part of the world that it is not always accompanied by a little of the matter of the first element.

Besides these two elements, I accept no others apart from a third, namely, the element of earth; I judge that its parts are bigger and move more slowly in comparison with parts of the second element, in the same way as the latter compare with those of the first element. I even think that it is enough to conceive of it as one or more large masses, the parts of which have very little motion, or none at all, which causes them to change position in relation to each other. If you find it surprising that I do not use qualities that are called 'heat', 'cold', 'humidity' and 'dryness' in order to explain these elements, as the philosophers do,[9] I shall tell you that these qualities seem to me to be in need of explanation themselves and, if I am not mistaken, that not only these four qualities but all the others too, and even all the forms of inanimate bodies, can be explained without having to assume for that purpose anything else in their matter apart from the motion, size, shape and arrangement of its parts. Because of this, I could easily explain to you why I do not accept any other elements apart from the three that I have described. For there should be a difference between them and the other bodies that philosophers call 'mixed' or blended and composite; this difference consists in the fact that the forms of such composite bodies always contain in themselves some qualities that are mutually incompatible and that negate each other or, at least, tend not to preserve each other, whereas the forms of elements should be simple and should not have any qualities that are not so perfectly compatible that each one tends towards the preservation of all the others.

Now I cannot find any such forms in the world apart from the three that I described. For that which I attributed to the first element consists in the fact that its parts move so extremely quickly and are so small that no other body is able to stop them and, in addition, they do not require any determinate size, shape or position. The form of the second element consists in the fact that its parts have a motion and size that are so moderate that, if there are many causes in the world which can increase their motion and decrease their size, there must be as many others which can do the exact opposite. Thus they remain permanently balanced in this same moderate condition. The form of the third element consists in the fact that its parts are so large, and so joined together, that they always have the force to resist the motion of other bodies.

Examine as much as you wish all the forms that could be given to mixed bodies by various movements, by different shapes and sizes, and by the various arrangements of the parts of matter; I am sure you will find none that does not have in itself qualities that tend to make it change and, in changing, to reduce to one of the forms of the elements.

For example, a flame – the form of which presupposes that it has parts that move very quickly, and also that they have some size, as was mentioned above – cannot exist for a long time without dying out. For, either the size of its parts gives them the force to act on other bodies, and that will cause a decrease in their movement; or the forcefulness of their motion will make them break apart when they collide with the bodies they encounter, and that will cause them to lose their size. Thus they could gradually reduce to the form of the third element or to that of the second, and some of them even to that of the first element. You can thereby recognize the difference between a flame or an ordinary fire, and the element of fire that I described. You should also know that the elements of air and earth, that is, the second and third elements, are likewise not similar to the gross air we breathe or to the earth on which we walk; but, in general, all the bodies that surround us are mixed or composite, and are subject to corruption.

However, it is not necessary for that reason to think that the elements have no places in the world that are especially destined for them and where they can be preserved perpetually in their natural purity. On the contrary, each part of matter always tends to be reduced to one of its forms and, once it is thus reduced, it never tends to leave that form again. Thus even if God had created only mixed bodies at the beginning, all bodies would still have had the opportunity to abandon their forms and to adopt those of the elements during the time since the world began. It is now very apparent therefore that all the bodies that are large enough to be counted among the most notable parts of the universe have the form of only one of the completely simple elements, and that there can be mixed bodies only on the surface of these larger bodies. But it is necessarily the case that there are some mixed bodies. Since the elements are naturally very incompatible, it is impossible for two of them to make contact without acting on each other on their surfaces and thereby giving the surface matter the different forms of such mixed bodies.

In this context, if we consider in general all the bodies of which the universe is composed, we shall find only three types which could be called large and could be included among its principal parts, viz. the sun and the fixed stars in the first type, the skies in the second, and the earth together with the planets and comets in the third. That is why we have good reason to think that the first element, completely uncontaminated, is the only form found in the sun and the fixed stars, that the skies have the form of the second element, and that the earth, together with the planets and comets, has that of the third.

I classify the planets and comets with the earth, because they resist

CHAPTER 5

the light as the earth does and they cause its rays to reflect, and therefore I find no difference between them. I also classify together the sun and fixed stars, and attribute to them a nature which is the complete opposite to that of the earth, for the action of their light alone is enough to inform me that their bodies are made of a very subtle and very mobile matter.

As regards the sky, in so far as it cannot be perceived by our senses, I think that there is reason to attribute to it a nature that is intermediate between that of luminous bodies, whose action we sense, and that of hard, heavy bodies, whose resistance we experience.

Finally, we perceive mixed bodies only on the surface of the earth; and if we consider that the whole space which contains them – i.e. all the space between the highest clouds and the deepest mines that have ever been dug for the extraction of metals, as a result of human avarice – is extremely small in comparison with the earth and with the immense expanses of the skies, we could easily imagine that all these mixed bodies together are only a crust that is produced on the surface of the earth by the action and mixture of the matter of the skies which surrounds it.

Thus we would have an opportunity to think that it is not only in the air that we breathe, but also in all other composite bodies, including the hardest stones and the heaviest metals, that there are parts of the element of air mixed with those of earth and also, consequently, some parts of the element of fire, because some of the latter are always found in the pores of the element of air.

But one must also acknowledge that, although there are parts of these three elements mixed together in all bodies, strictly speaking, it is only those that may be attributed to the third element, because of their size or the difficulty they have in moving, that compose all those that we see about us; for the parts of the other two elements are so subtle that they cannot be perceived by our senses. One could represent all these bodies as sponges, in which there are many pores or little holes which are always full of air, water or some other similar liquid; however, these liquids are not considered to be part of the composition of the sponge.

There are many things that remain for me to explain here, and I would even be quite happy to add some arguments to make my views more plausible. But in order for the length of this discourse to be less boring, I wish to cloak part of it in the invention of a fable through which, I hope, the truth will appear sufficiently and will be no less pleasing to see than if I present it completely naked.

CHAPTER 6

A Description of a New World, and the Qualities of the Matter of Which It Is Composed

Thus allow your thought to go outside this world, for a short time, to come to see a completely new world that I shall bring to life before you in imaginary spaces. The philosophers tell us that these spaces are infinite; they should be believed in this case, because they themselves invented them. But lest this infinity impede and burden us, let us not try to go to its limits; let us only go far enough so that we can lose sight of all the creatures that God created five or six thousand years ago. And when we have come to a stop there, in some determinate place, let us suppose that God creates anew all around us so much matter that, in whatever direction our imagination may be able to stretch, it would no longer perceive any place in it that is empty.

Although the sea is not empty, those who are on some vessel in the middle of it can extend their view, it seems, to infinity and there is still water beyond where they can see. Likewise, although our imagination seems capable of extending to infinity, and although this new matter is not supposed to be infinite, we may well suppose nevertheless that it fills spaces much larger than all those that we shall have imagined. And even, so that there may be nothing in all this which you find objectionable, let us not allow our imagination to wander as far as it can go; but let us restrain it intentionally to a determinate space which would be no greater, for example, than the distance between the earth

CHAPTER 6

and the principal stars of the firmament, and let us suppose that the matter that God will have created is extended far beyond that in all directions, for an indeterminate distance. For it is much more plausible, and we have much more power, to prescribe limits for the activity of our thought than to the works of God.

Now since we take the liberty to imagine this matter as we wish, we may attribute to it a nature in which there is nothing at all that cannot be known as perfectly as possible by everyone. For this reason, let us explicitly suppose that this matter does not have the form of earth, of fire, of air, or any other more specific form, such as that of wood, of a stone or of a metal, and that it does not have the properties of being hot or cold, light or heavy, nor any taste, odour, sound, colour, light or anything similar in the nature of which one could say that there is anything that is not clearly known by everyone.

On the other hand, let us also not think that this is the prime matter of the philosophers that has been so stripped of all forms and qualities that nothing remains in it which could be understood clearly. But let us conceive of it as a genuine body that is completely solid, that fills equally all the length, breadth and depth of this great space in the middle of which we have stopped our thought, so that each of its parts always occupies a part of this space which is so proportioned to its size that it would not fill a larger part nor squeeze itself into a smaller one, nor allow any other part of matter to occupy that place as long as it is there itself.

Let us add that this matter is divisible into all the parts and all the shapes that we can imagine, and that each of its parts can assume all the motions that we can conceive. Let us also suppose that God really divides it into many such parts, some of which are larger and others smaller, some with one shape and some with other shapes, as we wish to imagine them. It is not as if God separates these parts from each other in such a way that there is some vacuum between them; but let us think that the only distinction he introduces into these parts consists in the diversity of the motions he impresses on them, by making some of them begin to move in one direction and others in a different direction from the very first moment they were created, some of them faster and others slower (or not moving at all, if you wish), and by causing them to continue their motion subsequently in accordance with the ordinary laws of nature. For God has established these laws so wondrously that, even if we were to imagine that he created nothing more than what we have mentioned so far, and even if he imposed on it no order or proportion, but made it like the most confused and disordered chaos that poets could describe, the laws are enough to cause the parts of this chaos to disentangle from each other and to become arranged in such a good order that they would have the form of a very perfect world, in which one could see not only some light but also all the other things, both general and particular, that appear in the real world.

But before I explain this at greater length, pause to consider this chaos a little while longer and notice that it contains nothing that is not so perfectly known that you could not even pretend not to know it. For in the case of the qualities that I put in it, if you have noticed them, I have only supposed ones that you could imagine. And as regards the matter from which I composed it, there is nothing simpler or easier to know in inanimate creatures; and the idea of it is included so much in all the ideas that our imagination can form, that if you could not conceive of it you would never imagine anything.

None the less, since the philosophers are so subtle that they know how to find problems in things that seem to be extremely clear to other people, and since the memory of their prime matter – which they know is rather difficult to conceive of – could distract them from knowledge of the matter that I speak about, I must tell them at this point that, if I am not mistaken, all the difficulty they experience in their matter results only from the fact that they wish to distinguish it from its own quantity, and from its external extension, that is, from its property of occupying some space. However, I am willing to accept that they are correct about this, because I do not plan to stop to contradict them. But they should not find it strange, either, if I suppose that the quantity of the matter that I have described does not differ from its substance, no more than number differs from things that are numbered, and if I conceive of its extension, or of its property of occupying some space, not as an accident[10] but as its true form and essence. For they cannot deny that it is very easy to conceive of it in this way, and my plan is not, like theirs, to explain the things that are found in the real world but merely to imagine a world as I wish, in which there is nothing that the least subtle minds are unable to conceive and that, nevertheless, could not be created just as I imagined it.

If I had put the least thing in it which was obscure, it could happen that, within this obscurity, there might be some hidden contradiction that I would not have noticed, and thus, without thinking about it, I would suppose something impossible. Instead, by being able to imagine distinctly everything that I put into it, it is certain that, even if there were nothing like it in the old world, God can nevertheless create it in a new world; for it is certain that he can create all the things that we can imagine.

CHAPTER 7

The Laws of Nature of This New World

But I do not want to defer any longer telling you how nature, on its own, could disentangle the confusion of the chaos I spoke about, and what are the laws that God has imposed on it.

You should therefore realize, in the first place, that by 'nature' I do not understand here some goddess or any other kind of imaginary power, but that I use this word to mean matter itself in so far as I think of it with all the qualities that I attributed to it, all included together, and on condition that God continues to conserve it in the same way as he created it. For from the fact alone that he continues to conserve it in this way, it follows necessarily that there must be many changes in its parts; and since, properly speaking, they cannot be attributed to the action of God (since that does not change), I attribute them to nature, and the rules according to which these changes occur I call the laws of nature.

In order to understand this better, recall that, among the qualities of matter, we have supposed that its parts have had different motions since the beginning when they were created, and also that they are all in contact with each other on all sides without there being any vacuum between them. It follows necessarily from this that, once they began to move, they also began to change and diversify their motions by colliding with each other; and thus if God conserves them subsequently in the same way that he created them, he does not conserve them in the same state. That is to say, when God is always acting in the same way and is therefore always producing the same substantial effect, many changes occur in this effect as if by accident. It is easy to believe that God, who is immutable, as everyone must know, always acts in the same way. But without getting further involved in these metaphysical questions, I shall provide here two or three of the principal rules according to which one must think that God causes the nature of this new world to act and which will be enough, I believe, to let you know all the others.

The first rule is: that each individual part of matter always continues to be in the same state, as long as it is not forced to change that state by collision with others. That is, if it has a certain size, it will never become smaller unless other parts of matter divide it; if it is round or square, it will never change its shape unless others force it to do so; if it is stationary in some place, it will never leave it unless others drive it out; and if it ever begins to move, it will always continue to move with an equal force until others stop it or slow it down.

Everyone believes that this same rule is observed in the old world with respect to size, shape, rest and a thousand other similar things. But the philosophers have exempted motion from this rule, although that is the thing that I most expressly wish to include in it. Do not think, for that reason, that I plan to contradict them; the motion of which they speak is so very different from that which I conceive there, that it could easily happen that what is true of one of them is not true of the other.

They themselves concede that very little is known about the nature of their motion, and in order to make it intelligible in some way, they have not been able so far to explain it more clearly than in these terms: *'motus est actus entis in potentia, prout in potentia est'*.[11] These are such obscure terms for me that I am forced to leave them here in their

CHAPTER 7

original language, because I would not be able to interpret them. (And, indeed, these words: 'motion is the act of a being in potency, in so far as it is in potency', are not any clearer because they are translated.) In contrast, the nature of the motion that I wish to speak about here is so easy to know that even geometers, who of all people are most anxious to conceive very clearly the things that they study, have thought it was simpler and more intelligible than the nature of their surfaces and their lines. This is clear from the fact that they have explained the line as the motion of a point, and a surface by the motion of a line.

The philosophers also suppose many motions which, they think, can be realized without any body changing its place, such as those that they call '*motus ad formam*', '*motus ad calorem*', '*motus ad quantitatem*' [change of form, change of heat and change of quantity], and a thousand others. As far as I am concerned, I know of no motion apart from that which is easier to conceive than the lines of geometers, which makes bodies move from one place to another and occupy successively all the intervening spaces.

In addition, they attribute to the least of these motions a being which is much more solid and real than they do to rest; the latter, they say, is only the privation of motion. But I think that rest is just as much a quality that must be attributed to matter while it remains in the same place, as motion is one that should be attributed to it when it is changing place.

Finally, the motion of which they speak is of such a strange nature that, whereas all other things have as their end their own perfection and strive only to preserve themselves, motion has no other end or goal apart from rest and, contrary to all the laws of nature, it tries to destroy itself. In contrast, the motion that I suppose follows the same laws of nature as do generally all the dispositions and qualities that occur in matter, both those that the learned call '*modos & entia rationis cum fundamento in re*' (modes, and beings of reason which have a foundation in reality), and those they call '*qualitates reales*' (their real qualities) in which, I frankly confess, I cannot find any more reality than in the others.

I hypothesize as a second rule: when one body pushes another, it could not give the other any motion except by simultaneously losing as much of its own motion, nor could it take away any of the other's motion unless its own motion increases by the same amount. This rule, together with the preceding one, corresponds very well to all the experiences in which we see that a body begins or ceases to move because it is pushed or stopped by some other body. Having assumed the preceding rule, we are free from the difficulty in which the learned find themselves, when they wish to explain the fact that a stone continues to move for some time after leaving the hand of whoever threw it; for we could be asked, instead, why it does not continue to move for ever? But it is easy to give the reason why: for who can deny that the air in which it moves offers it some resistance? It can be heard whistling as it cuts through the air, and if one moves a fan in the air, or some other very light and very extended body, one can even feel in the weight of one's hand that the air impedes its motion; this is very far from keeping it moving, as some wish to say.[12] But if one fails to explain the effect of the air's resistance in accordance with our second rule, and if one thinks that the more a body can resist the more it is able to stop the motion of others, as one might initially be able to convince oneself, one would then have great difficulty in explaining why the motion of a stone is reduced more in colliding with a soft body – the resistance of which is moderate – than when it collides with a hard body which resists it more. Likewise, as soon as it has made a little effort against the latter, why is it that it does an immediate about-face, as it were, rather than stop or interrupt the motion of its subject? However, if we suppose this rule, there is no difficulty at all in this. For it teaches us that the motion of one body is not retarded by collision with another in proportion to the latter's resistance to it, but only in proportion to overcoming its resistance; and to the extent that it obeys the other body, it receives in itself the force to move that the other body loses.[13]

Now although in most of the motions that we see in the real world we cannot perceive that the bodies which begin or cease to move are pushed or stopped by others, that does not give us any reason to decide that these two rules are not observed exactly there. For it is certain that these bodies can often receive their motion from the two elements of fire and air, which are always present within them without being capable of being perceived there (as has just been said), or they may even receive motion from the gross air, which cannot be perceived either. They can also transfer this motion sometimes to this gross air, and sometimes to the whole mass of the earth, where, once it is dispersed, it cannot be perceived either.

But even if everything that our senses have ever experienced in the real world seemed manifestly to be incompatible with what is contained in these two rules, the reasoning that taught them to me seems to be so convincing, that I would still think I had to hypothesize them in the new world that I am describing for you. For what more firm and more solid foundation could one find to establish a truth, even if one wished to choose it at will, than the very firmness and immutability which is in God?

Now these two rules follow manifestly from this alone, that God is immutable and, since he always acts in the same way, he always produces the same effect. For if we suppose that he placed a certain quantity of motion in the whole of matter in general from the first moment that he created it, one must accept either that he always conserves the same amount of it there or else not believe that he always acts in the same way. And if we also suppose that, from this first moment, the various parts of matter in which these motions were unequally distributed began to retain them or to transfer them from one to another, in so far as they may have had the force to do so, one must necessarily think that he makes them continue the same process for ever. That is what is contained in these two rules.

I would add as a third rule: when a body moves, even if its motion were to occur most often along a curved line, and if it could never move except in a way which is in some sense circular (as was said above), nevertheless each of its parts individually always tends to continue its motion in a straight line. And thus their action – that is, the inclination of these parts to move – is distinct from their motion.

For example, if a wheel is turned on its axis, although all its parts go in a circle – because, since they are joined with each other, they cannot go elsewhere – nevertheless their inclination is to go in a straight line. This can be seen clearly if by chance one of them becomes detached from the others; for as soon as it is free, its motion ceases to be circular and it continues in a straight line.

Likewise, when one swings a stone in a sling, not only does it go perfectly straight as soon as it is released, but also, during the whole time that it is in the sling, it presses against the fold in the sling and causes the cord to stretch.[14] This shows clearly that it always has a tendency to go in a straight line, and that it goes around in a circle only by being constrained.

This rule is based on the same foundation as the other two, and depends only on the fact that God conserves each thing by a continuous action, and therefore he does not conserve it as it may have been some time previously, but precisely as it is at the very instant that he conserves it. Now, among all the motions, only motion in a straight line is completely simple and its whole nature is contained in one instant. For, in order to conceive of it, it is enough to think that a body is in the act of moving in a certain direction, and this is present in each of the moments that can be determined during the time it is moving. In contrast, in order to conceive of a circular motion, or any other possible motion, one must consider minimally two of its moments, or preferably two parts of the motion and the relationship between them. But lest the philosophers, or rather the sophists, seize the opportunity here to apply their redundant subtleties, notice that I do not say that rectilinear motion can occur in an instant; all I say is that everything required to produce it is present in bodies in every instant that can be determined during the time they are moving, whereas everything required to produce circular motion is not similarly present.

For example, if a stone is moving in a sling along the circle marked AB [in the figure below], and if you were to consider it exactly as it is at the instant when it arrives at point A, you would surely find that it is in the process of moving, because it does not stop there, and that it is moving in a certain direction, that is, towards C, because that is the way in which its action is determined at that moment. But you could not find anything there that makes its motion circular. Thus if we suppose that it begins to leave the sling at that moment, and that God continues to conserve it in the state it is in at that moment, it is

CHAPTER 7

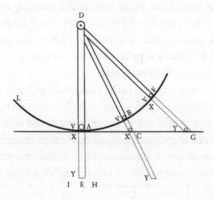

certain that he will not conserve it with an inclination to move in a circle along the line *AB*, but with an inclination to go straight towards point *C*.

Thus, according to this rule, one must say that God alone is the author of all the motions that occur in the world, in so far as they exist and in so far as they are rectilinear, but that it is the various dispositions of mattter that make them irregular and curved, just as theologians teach us that God is also the author of all our actions, in so far as they exist and in so far as they have some goodness, but it is the various dispositions of our wills that can make them evil.

I could provide many rules here to determine in detail when, and how, and how much, the motion of each body may be diverted, and increased or decreased, by colliding with other bodies, which include in summary all the effects of nature.[15] But I am satisfied to inform you that, apart from the three rules that I have explained, I do not wish to hypothesize any others apart from those that follow infallibly from the eternal truths on which mathematicians have been accustomed to build their most certain and evident demonstrations – the truths, I say, according to which God himself has taught us that he arranged all things in number, weight and size, and the knowledge of which is so natural to our souls that we could not fail to think they are infallible as long as we conceive of them distinctly. Nor could we doubt that, if God had created many worlds, they would not be as true in all of them as in this one. Thus those who could examine sufficiently the consequences of these truths and of our rules, could be able to discover effects by their causes, and, to explain myself in the language of the schools, they could have a priori demonstrations of everything that could be produced in this new world.

To avoid allowing any exception that could prevent this, we shall add to our hypotheses, if you wish, that God will never perform any miracle in this new world and that intellects, or the rational souls that we may later suppose are present there, will never disturb the ordinary course of nature in any way. However, I do not promise to provide you with exact demonstrations of everything that I say from here on; it will be enough if I open the path for you by which you can find them yourself when you take the trouble to look for them. Most minds lose interest when things are made too easy for them; and I have to use both shade and bright colours to make a pleasant painting for you here. Thus I shall be content to continue the description that I have begun, as if my only plan were to tell you a fable.

Rules for Guiding One's Intelligence in Searching for the Truth

(1628; published 1701)

NOTE ON THE TEXT

This is an early version of Descartes's method which was composed intermittently between 1619 and 1628. Work on redrafting the text seems to have been suspended about 1628, and the manuscript remained unpublished during the author's lifetime. After Descartes's death in 1650, the manuscript was passed to Claude Clerselier, his literary executor, who arranged for publication of *The World* and three volumes of correspondence. However, Clerselier died in 1684, prior to publishing the *Rules*, but not before showing the manuscript to various other philosophers who were interested in Cartesian philosophy. These included A. Arnauld and P. Nicole, the authors of the *Port-Royal Logic*; Descartes's biographer, Baillet; and Nicolas Poisson, who wrote his own introduction to method in the sciences. Some indications of the contents of the original manuscript, and translations of short excerpts, were published in French by these authors during the seventeenth century; but the original manuscript was lost. A Dutch translation of the *Rules* was published in 1684 (usually identified as **N**), and the first Latin edition appeared in Amsterdam in 1701 (denoted as **A**). A copy of the manuscript prepared for Leibniz was found among his papers at the Hanover library in the nineteenth century (denoted as **H**).

The original plan of the work was to provide thirty-six rules, grouped in three sets of twelve each. Only twenty-one rules survive, the last three of which are incomplete. This translation is based on the critical edition prepared by G. Crapulli, *Regulae ad directionem ingenii* (Nijhoff, The Hague, 1966), whose text was based on **N**, **A** and **H**. In preparing this edition I also consulted the French translation of Jean-Luc Marion, *Règles utiles et claires pour la direction de l'esprit et la recherche de la vérité* (Nijhoff, The Hague, 1977).

Since the original manuscript was lost, we are not sure what working title Descartes had given to this draft essay. The inventory of manuscripts found after Descartes's death in Stockholm identified the *Rules* as: '*Un traité des règles utiles et claires pour la direction de l'Esprit en la recherche de la Vérité*'. In fact, all references to the title of this work, with the exception of that used in edition **A**, include some mention of searching for the truth; however, it has been customary since its original publication to repeat the short title used in **A**, *Regulae ad directionem ingenii*, even though it is evidently not authoritative. I have followed Jean-Luc Marion's suggestion by restoring a reference to 'searching for the truth' in the title, but have adopted the slightly shorter title used by Baillet.

Since Descartes left this manuscript unfinished, one could hardly argue that his choice of where to place new paragraphs was philosophically significant. In the interests of readability, therefore, I have supplied extra indentions wherever they seemed appropriate.

RULES FOR GUIDING ONE'S INTELLIGENCE IN SEARCHING FOR THE TRUTH

RULE ONE

The aim of studies should be to guide one's intelligence towards making well-founded, true judgements about everything that one encounters.

Whenever people notice some similarity between two things, they habitually judge that whatever they find to be true of either one of them applies to both of them, even in the case of some feature with respect to which they differ. Thus they mistakenly compare the sciences, which consist exclusively of knowledge by the mind, with the arts, which presuppose some kind of training and skill on the part of the body. And when they notice that all the arts cannot be learned simultaneously by the same person, and that one becomes a very skilled artisan more easily by exercising only one skill – because the same hands are not as adaptable to cultivating the fields and playing the zither, or to many different skills like those, as they would be to only one of them – they imagine that the same thing is true in the sciences. They distinguish them from each other by the diversity of their objects, and then think that they should each be sought separately by omitting all the others. But they are completely wrong about this. For since all the sciences are nothing other than human wisdom, which remains always one and the same even when it is applied to different subjects, and since human wisdom is no more changed than is the light of the sun by the variety of things that it illuminates, there is no need to constrain our intelligence within such limits.

Nor does knowledge of one truth prevent us from discovering another, as happens in the practice of an art; on the contrary, it helps us. Indeed it seems surprising to me that many people investigate very diligently the powers of plants, the motions of the stars, the transmutation of metals, and the objects of similar disciplines, but that almost no one thinks about common sense[1] or this universal wisdom, even though none of the other sciences are valuable except in so far as they contribute in some way to this. Therefore it is right that we propose this as the first rule of all, because there is nothing that will lead us astray more readily from the right path by which to search for truth, than if we direct ourselves towards certain particular studies and not towards this general objective. I am not referring to various immoral or blameworthy objectives, such as vainglory or filthy lucre; it is obvious that specious reasons, and illusions that are accommodated to the lowest minds, provide a much more direct path to them than could reliable knowledge of the truth. I am speaking instead of honourable and praiseworthy objectives, because we are often deceived by them more subtly – for example, if we seek sciences that are useful to human needs, or the pleasure that is found in contemplation of the truth and is almost the only happiness in this life that is complete and untroubled by sorrow. We can expect these as the legitimate fruits of the sciences; but if we include them among the things that should be studied, they often cause us to omit many things that are necessary for knowing other things, because the latter seem initially to be either less useful or less interesting. But it must be accepted that all the sciences are so mutually interconnected that it would be much easier to learn them all together than to separate one of them from the others.

Therefore if someone wishes seriously to investigate the truth about things, they should not choose some particular science, because all the sciences are interconnected and dependent on each other. They should think instead only about increasing the natural light of reason, not in order to resolve this or that problem of scholastic philosophy, but in order for their intellect to instruct their will about what choice to make in each of life's decisions; and in a short time they will be surprised that they have made much more progress than those who study particular things, and thus they will have achieved not only all the things that others desire, but they will also have achieved much more than others could hope for.

RULE TWO

We should be concerned only with those objects, for which our intelligence seems adequate to achieve a certain and indubitable knowledge.

Every science[2] is certain and evident knowledge. Someone who doubts about many things is no wiser than someone else who has never thought about them; but they seem none the less to be less wise than the other person, if they have formed a false opinion about something. Thus it is better never to study than to be concerned with things that are so difficult that, since we are unable to distinguish true and false opinions about them, we are forced to accept what is doubtful as if it were certain; for in questions like this there is a greater danger of decreasing our learning than a hope of increasing it. Thus by means of this proposal we reject all knowledge that is merely probable, and we decide to believe only what is known perfectly and cannot be doubted. Although educated people may convince themselves that there are very few such items of knowledge because, due to a common human weakness, they have failed to reflect on them and think that they are too easy and too obvious to everyone; but I claim that they are much more numerous than is believed, and that such truths are enough to demonstrate with certainty innumerable propositions about which, so far, it has been possible to write only in a probabilistic fashion. And since they thought it was beneath the dignity of an educated person to admit that there was something they did not know, they became so used to embellishing their false reasons that they later convinced themselves of them and thus defended them as if they were true.

However, if we observe this rule strictly, there will be very few things to the study of which we can devote ourselves. For there is hardly any question in the sciences about which intelligent scholars have not often disagreed. When two people make contrary judgements about the same thing, however, it is certain that at least one of them is mistaken, and it seems as if neither one of them has scientific knowledge. For if either one had reasons that were clear and certain, they could present them to the other in such a way that their intellect would eventually be convinced. Thus it seems that we cannot acquire perfect knowledge of probable opinions like this, because it would be foolish to hope that we could make more progress on them than others have made. Thus, if we are assessing the situation correctly, the observance of this rule reduces us to arithmetic and geometry, which alone remain among all the sciences discovered so far.

It does not follow from this that we condemn the type of philosophizing that others have previously discovered, or those missiles – the probable arguments of the scholastics – which are very suitable for belligerent disputes. They exercise the intelligence of young people and develop in them a certain rivalry; and it is far better for them to be informed about such views, even if they are obviously uncertain, since they are disputed among the learned, than to be left completely to their own devices. Without any guide, they might wander towards precipices; but as long as they follow in their teachers' footsteps, then, even if they sometimes turn aside from the truth, they will surely find a path that is more secure, at least in the sense that it has already been tested by prudent people. We ourselves are glad that we were once taught in this way in the schools; but since we are now released from the oath that bound us to the words of a teacher and have eventually become mature enough to guide ourselves,[3] if we wish seriously to propose rules for ourselves by the use of which we shall reach the heights of human knowledge, the following should clearly be included among the first rules: to take care not to waste our free time, as many people do who neglect everything that is easy and who are concerned only with difficult things, about which they ingeniously construct hypotheses that are definitely very subtle and arguments that are very probable. However after much hard work they eventually realize that they have merely increased the number of their doubts, but have acquired no scientific knowledge.

Now that we have said above that, among all the disciplines known to others, arithmetic and geometry alone are free from every taint of falsehood or uncertainty, in order to explain more carefully why this is so we should note that we can arrive at knowledge of things by means of two paths, viz. by experience or deduction. It should also be noted that experiences of things are often deceptive, whereas a deduction – or the pure inference of one thing from another – may be overlooked if it is not apparent, but it can never be performed badly by a minimally rational intellect. The chains of dialecticians, by which they think they can regulate human reason, seem to me to be of little use for deduction, although I do not deny that they are very appropriate for other purposes. No deception that can occur to human beings (I am not claiming this applies to animals) ever results from a poor inference, but only from the fact that various experiences that are poorly understood are accepted, or rash judgements are made without any foundation.

It is clear from this why arithmetic and geometry are much more certain than other disciplines. The reason is that these alone are concerned with an object that is so pure and simple that they evidently presuppose nothing that experience might render uncertain, but they consist exclusively of conclusions that are deduced by reason. They are therefore the easiest and clearest of all disciplines, and they have the kind of object that we require since it seems that, in their case, it is scarcely possible for someone to be mistaken except by not paying attention. It should not be surprising, therefore, if many people spontaneously apply their intelligence, instead, to other arts or to philosophy; this happens because everyone is more confident in allowing themselves the freedom to guess about obscure matters than about what is clear, and it is much easier to make a guess about some question than to arrive at the truth itself, no matter how easy it may be.

One should conclude from all this, not that arithmetic and geometry alone should be studied, but only that, in seeking the right path to the truth, one should not be concerned with any object about which one cannot have as much certainty as in the demonstrations of arithmetic and geometry.

RULE THREE

We should seek whatever we can intuit clearly and evidently or what we can deduce with certainty about any proposed objects, and not what others have thought about them or what we ourselves might guess; for scientific knowledge cannot be acquired in any other way.

The books of the ancients should be read because it is very beneficial for us to be able to use the work of so many people, both to learn what has already been correctly discovered and also to be informed about what remains to be thought out in all disciplines. Meanwhile there is a great danger that the stains of errors that could be acquired from too close a reading of those books would stick to us despite ourselves and despite the care we take. For writers usually have an intelligence such that, once they have fallen for some controversial view because of their incautious credulity, they always try to draw us into the same view by very subtle reasons. On the other hand, whenever they are lucky enough to discover something certain and evident, they never reveal it unless it is wrapped up in various obscurities, for fear that the simplicity of their argument would diminish the importance of the discovery or because they begrudge us the bare truth.

But even if all writers were sincere and open-minded, and never tried to pass off anything doubtful as if it were true, but revealed everything to us in good faith, we would always be uncertain about who to believe because there is hardly anything claimed by one without someone else asserting the opposite. There would also be no point in counting votes, in order to follow the view that has most supporters. For if the question at stake is difficult, it is much more likely that the truth about it could be discovered by few people rather than by many. Even if everyone agreed about something, their teaching would still not be enough. For example, we will never become mathematicians, even if we remember all the demonstrations of others, unless we are also able to use our intelligence to solve whatever problems we encounter. Nor shall we ever become philosophers by reading all the arguments of Plato and Aristotle, if we are unable to make a definite judgement about questions that are raised, for by doing this we would seem to have learned only histories rather than sciences.

We are also advised that no conjectures should ever be mixed in with the judgements we make about the truth of things. This realization is not insignificant. The main reason why the common philosophy contains nothing that is evident and certain enough that it cannot be challenged is that students are, first, not content to acknowledge clear and certain things, but they dare to make claims about obscure and unknown things, which they have been able to reach only by means of probable conjectures. They then gradually put their whole trust in such claims and, by mixing them indiscriminately with true and evident beliefs, they were eventually unable to conclude anything from them which did not seem to depend on one of the probable propositions, and which was not therefore uncertain.

But lest we fall into the same mistake in future, we shall list here all the actions of our intellect, by which we can arrive at knowledge of things without any fear of error. We accept only two, namely, intuition and deduction.

By 'intuition' I understand, not the changing testimony of the senses or the false judgement of an imagination when it composes images badly, but the conception of a pure and attentive mind that is so easy and so distinct that no doubt remains subsequently about what we understand; or, what is the same thing, the undoubting conception of a pure and attentive mind which arises from the light of reason alone, and which is more certain than deduction because it is simpler – though we noted above that even human beings cannot perform a deduction poorly. Thus each person can mentally intuit that they exist, that they think, that a triangle is bound by only three sides, that a globe has a single surface, and similar things which are much more numerous than most people realize, because they think it is below their dignity to turn their minds to such easy things.

However, if some people are disturbed by this novel use of the word 'intuition', and of other words that I am forced to change from their usual meaning in the following pages, I hereby advise them as a general point that I am not thinking at all about the way in which certain words have been used in recent times in the schools, for it would be very difficult to use those same words and to understand them in a completely different way. All I do is to notice what particular words mean in Latin, so that, whenever I lack appropriate words, I shall transfer to my own meaning whatever words seem most suitable.

However, this certainty and evidence of intuition is required, not only for individual propositions, but also for any discourse. For example, the following conclusion is drawn: two plus two is equal to three plus one. One must intuit, not only that two plus two equals four and that three plus one also equals four, but that the third proposition follows necessarily from these two.

There may therefore be some doubt why, in addition to intuition, we have added another form of knowing, namely, knowing by deduction, by which we understand everything that follows necessarily from other things that are known with certainty. But this had to be done in this way because many things are known with certainty even though they themselves are not evident, on condition only that they are deduced from principles that are true and certain, by means of a continuous and uninterrrupted movement of thought which intuits each element clearly. This is similar to the way in which, in the case of a very long chain, we know that the last ring is connected to the first, even though we do not comprehend by one and the same visual intuition all the intermediate rings on which that connection depends, provided we survey them one after another and remember that each individual ring, from the first to the last, is joined with the one next to it. Therefore we distinguish here a mental intuition from a deduction which is certain, because we conceive of a motion or some kind of succession in the latter, but not in the former. Moreover, present evidence is not necessary for deduction, as it is in the case of intuition; instead, it borrows its certainty in some way from memory. It follows that those propositions that are inferred immediately from first principles can be said, from different points of view, to be known in one sense by intuition and in another sense by deduction. But the first principles themselves are known only by intuition and, in contrast, the remote conclusions are known only by deduction.

These two paths to scientific knowledge are very certain and, as far as our intelligence is concerned, no addional ones should be conceded, and all others should be rejected as suspect and liable to errors. However that does not prevent us from believing that everything that was revealed by God is more certain than all knowledge, since faith in these things, however obscure they may be, is not an act of intelligence but of will. Besides, if the latter had foundations in the intellect, they should and could be discovered, more than anything else, by either of the two paths just mentioned, as we shall perhaps show elsewhere.[4]

RULE FOUR

A method is required in order to search for the truth about things.

Mortals are so bound by blind curiosity that they often lead their intelligence down unfamiliar paths without any reason for hope, but merely to test if what they seek may happen to lie there. It is as if someone were so consumed by a foolish desire to find treasure, that they constantly wandered the streets hoping to find, by chance, something that had been lost by a passer-by. This is how almost all chemists, most geometers and quite a number of philosophers study. I certainly do not deny that they sometimes have such luck in their wandering that they find something true; however, I concede in that case that they are lucky rather than diligent. But it is much more satisfactory never to think about seeking the truth about anything, than to do so without a method, because it is very certain that the natural light is obscured and our intelligence is blinded by such disordered studies and obscure meditations. Anyone who gets used to walking thus in the shadows weakens their eyesight to such an extent that they cannot subsequently tolerate daylight. This is confirmed by experience, for we see very often that those who have never studied judge much more reliably and clearly about simple things than those who have spent all their time in the schools. By a 'method', however, I understand easy and certain rules such that, if anyone were to use them carefully, they would never accept what is false as true and, without wasting their mental effort but always increasing their scientific knowledge gradually, they would arrive at a true knowledge of all the things that they are capable of knowing.

It should be noted that there are two parts here: not to accept anything false as true, and to arrive at knowledge of everything. For if there is something we do not know among all the things that we are capable of knowing, that happens only because we have not noticed any path that would lead us to such knowledge, or because we have fallen into the opposite mistake. But if a method explains properly how mental intuition should be used, so that we do not fall into error (which is the opposite of the truth), and how deductions should be found so that we come to have knowledge of all things; it seems to me that nothing further is required to make it complete, since the only way to acquire scientific knowledge is by intuition and deduction, as was said above. A method cannot be extended to teach us how these operations themselves should be performed, since they are the simplest of all and are primary. Thus unless our intellect were already able to make use of them, it would not comprehend any rules of a method, no matter how easy they were. Other mental operations, which dialectic claims to direct with the assistance of these primary ones, are useless in this context or, rather, they should be classified as impediments, because nothing can be added to the pure light of reason that would not obscure it in some way or other.

Since this method is so useful that, without it, it would seem to be more harmful than beneficial to take on the work involved in study, I easily convince myself that it was already noticed in some way by the great minds of the past, or that they were guided to it by nature alone. For the human mind has a divine I-know-not-what, in which the first seeds of useful thoughts have been sown in such a way that, oftentimes, despite being neglected and suffocated by obstructive studies, they produce spontaneous fruit. We experience this in the easiest sciences, arithmetic and geometry; for we recognize sufficiently that the ancient geometers used some kind of analysis which they applied to the resolution of all problems, although they did not pass it on to their successors. And a certain kind of arithmetic is already thriving, which they call 'algebra', and they have made it as successful in numbers as the ancients did in respect of geometrical shapes. But these two are nothing more than the spontaneous fruits that have resulted from the innate principles of this method; it is not surprising that they have thus far flourished more when applied to the very simple objects of these disciplines than in the case of others, in which greater impediments usually suffocate them. But even in the latter, there is no doubt that they could achieve perfect maturity provided they were very carefully cultivated.

This is indeed what I have principally undertaken to do in this treatise. I would also not think that these rules were significant if they were enough only to resolve the inane problems by which logicians and geometers have become accustomed to wasting their time; in that case

I would think that I may have achieved nothing more than to have dealt with trifles more subtly than others. Although I am about to say much about figures and numbers here, because it is impossible to look for examples that are as evident and certain in other disciplines, whoever pays attention to my meaning will easily recognize that I am not at all thinking about ordinary mathematics here, but that I am expounding a new discipline of which mathematics is the outer layer rather than its parts. This discipline must contain the primary elements of human reason, and must extend to eliciting truths from any subject. To speak freely, I am convinced that it is more powerful than any other human knowledge that we have inherited, since it is the source of all other knowledge. I have spoken about an 'outer layer' – not in the sense that I want to hide this doctrine and surround it, so that ordinary people are kept at a distance from it, but in the sense, rather, of clothing and adorning it in such a way that it is better adapted to human intelligence.

When I first applied my mind to mathematical disciplines, I immediately read through most of what mathematical authors usually teach us, and I especially cultivated arithmetic and geometry, because they were said to be the most simple and to be like paths to the other [branches of mathematics]. But there were no writers in either of them, among those that I happened to come across, who satisfied me fully. For I read many things in them about numbers, that I found were true once I went through the calculations myself; and with respect to geometrical shapes, they in some sense revealed many before my very eyes, and they drew some inferences from them. But they did not seem to show the mind adequately why things were as they were, and how they were discovered. It was not surprising, therefore, that even very learned and intelligent people abandoned those disciplines or soon neglected them as childish and vain or, on the other hand, were discouraged at the outset from learning them because they were too difficult and complicated. For there is nothing more foolish than to be so concerned with bare numbers and imaginary figures that we seem to wish to remain content with knowledge of such trifles, and to devote ourselves to such superficial demonstrations that – since they are discovered by chance more often than by skill, and pertain to the eyes and the imagination more than to the intellect – we become unaccustomed in some way to using reason itself. At the same time there is nothing more complicated than to generate new difficulties that are intertwined with confused numbers by using such a style of proving.

When I later thought why it was that the first inventors of philosophy were unwilling to admit anyone to the study of wisdom who lacked expertise in mathesis, as if this discipline appeared to them to be the simplest and most necessary for training and preparing minds to grasp other more important sciences, I wondered whether they were familiar with a mathesis that is very different from the one which is common in our age. It is not that I think they knew it perfectly, for their unreasonable celebrations and sacrifices on the occasion of trivial discoveries show clearly that they were unsophisticated. Nor am I convinced otherwise by some of their machines, which are celebrated by historians. They may have been very simple, and may easily have been raised to the status of miracles by an ignorant and impressionable multitude. However, I am convinced that certain seeds of truth which are innate in the human mind — and which we extinguish daily in ourselves by the number of errors we read and hear — had such strength in that primitive and pure antiquity, that the same light of nature by which they saw that virtue is preferable to pleasure and duty to utility (even if they did not know why) made them recognize true ideas in philosophy and mathesis, although they were not yet able to pursue those sciences perfectly. Indeed, some traces of this true mathesis seem to me to appear in Pappus and Diophantes,[5] who, although they were not in the first age, lived many centuries before our time. I would almost believe that, by a pernicious cunning, this was suppressed by these writers themselves. For in the same way that many discoveries were concealed by their inventors, they may have feared, because it was very easy and simple, that it would be lost if it were revealed, and they preferred to reveal instead, as the results of their work, certain sterile truths that are demonstrated by subtle arguments so that we would admire them, rather than teach us the art itself that would have dispelled our admiration completely. There were eventually some very gifted men who tried to revive it in this century; for the art that they call by the barbaric name of 'algebra' seems to me to be identical with it, if only it can be divested of the multiplicity of numbers and inexplicable figures by which it is camouflaged, so that it would no longer lack the greatest clarity and simplicity that we assume ought to be found in genuine mathesis.

When these thoughts recalled me from the particular study of arithmetic and geometry to searching for a certain general mathesis, I first inquired what precisely everyone understood by that term, and why not only the disciplines already mentioned but also astronomy, music, optics, mechanics and many others are also said to be parts of mathematics. It is not enough here to examine the origin of the term; since the word 'mathesis' means the same as 'discipline', the other disciplines [i.e. astronomy, etc.] have as much right as geometry itself to be called mathematics. But we see that there is almost no one with the slightest education who does not distinguish easily, among the things they encounter, between what pertains to mathematics and what pertains to the other disciplines. It became clear eventually to anyone who examined it more closely that only those things in which some order or measure is examined pertain to mathesis, and it is irrelevant whether such a measure is sought in numbers, or figures, or stars, or sounds or any other object.

Therefore there must be some general science which explains everything that can be learned about order and measure, which is not confined to any particular subject matter, and is called universal mathesis. This is not a borrowed name, but one with a long and accepted usage; for it includes everything on account of which other sciences are called parts of mathematics. The extent to which it is superior in usefulness and facility to these subordinate sciences is clear from the fact that it applies to everything to which they are applied, and to more besides; and if it includes some difficulties, they are also found in the other disciplines, whereas the latter include other difficulties that result from their specific subject matter and that are not found in universal mathesis. Now, since everyone knows its name and understands what it is concerned with, even if they do not study it, how does it happen that many people laboriously pursue the other disciplines that depend on it, while no one bothers to learn this discipline itself? I would certainly wonder about that if I had not been aware, for a long time, that the human intelligence always bypasses what it thinks it can learn easily and hurries headlong towards novelties that are more sublime.

But, conscious of my weakness, I decided that I would observe stubbornly an order in seeking knowledge of things so that, always beginning from the most simple and easy things, I would never proceed to others until it seemed that I could hope for nothing more from them. That is why I have cultivated up to now this universal mathesis as much as I could, so that, from now on, I think I can study the slightly more obscure sciences, without being premature in my application. But before I set out from here, I shall try to collect together and arrange in order whatever I perceived worth noting from my previous studies so that, when my memory is dimmed by increasing age, I can find it all easily in this little book if the need arises, and also, having unburdened my memory of them, I can devote my mind more freely to other things.

RULE FIVE

The entire method consists in the order and arrangement of those things to which the mind's eye must turn so that we can discover some truth. But we shall observe this method exactly if we reduce convoluted and obscure propositions step by step to more simple ones, and if we then try to ascend by the same steps to knowledge of all the others, beginning from an intuition of all the most simple propositions.

This contains, on its own, the sum of all human diligence and this rule must be observed by anyone who is searching for knowledge of things, just as they would follow the thread of Theseus to enter the labyrinth. But many people do not reflect on what it prescribes, ignore it completely, or presume that they do not need it, and they frequently examine the most difficult questions in such a disorderly manner that they seem to me to act as if they are trying to go from the bottom of some building to its top in one step, either by bypassing the steps of the stairs which are designed for that purpose or by not noticing them at all. All astrologers do this when, not knowing the nature of the skies and not having even observed their motions accurately, they hope to be able to say what their effects are. Most people who study mechanics without physics do likewise, when they casually construct new instruments for producing motions. Philosophers are the same who, neglecting experience, think that the truth will spring from their own brains as Minerva did from the head of Jove.

Indeed, they all clearly break this rule. But because the order required in this context is often so obscure and complicated that everyone is not able to recognize what it is like, it is hardly possible for them to take enough care not to go astray unless they observe diligently what is explained in the following proposition.

RULE SIX

In order to distinguish the simplest things from those that are complex and to search for them in an orderly way, one should notice what is most simple in each sequence of things in which we have directly deduced some truths from others, and how all the others are more, or less, or equally distant from the most simple item.

Although this proposition seems to teach us nothing very novel, it still contains the principal secret of the art and there is none more useful in this whole treatise. For it advises us that all things can be arranged in various sequences, not indeed in so far as they are referred to some metaphysical class,[6] as the philosophers have divided them into their categories, but in so far as some can be known from others; thus whenever some difficulty arises, we can immediately notice whether it would be worth while to examine some others first, which ones should be examined first, and in what order.

In order to do this properly, however, it should be noted first that, in the sense in which things can be useful for our project – where we do not examine their natures in isolation but compare them with each other, so that we can know some of them from others – all things can be said to be either absolute or relative.

I apply the term 'absolute' to anything that contains in itself the pure and simple nature that is in question – thus everything that is thought to be independent, a cause, simple, universal, one, equal, similar, straight and other things like that. I also call the first thing the most simple and easy, so that we may use it to resolve questions.

In contrast, something is said to be relative when it participates in the same nature or, at least, in something from the same nature, and accordingly can be referred to what is absolute and can be deduced from it through some sequence. But the concept of the relative also includes other things, which I call 'relations'. These include anything that is said to be dependent, an effect, composite, particular, many, unequal, dissimilar, oblique, etc. 'Relatives' are further removed from 'absolutes' in so far as they contain more relations of this kind which are mutually subordinate to each other. We are advised by this rule to distinguish all these and to notice the mutual connection and the natural order among them so that, beginning from the last one, we can arrive at that which is most absolute by passing through all the others.

The secret of the whole art consists in taking careful notice of what is most absolute among all things. For some are indeed more absolute than others from one point of view, but from a different perspective they are more relative. Thus what is universal is more absolute than what is particular, because it has a simpler nature; but it can also be said to be more relative than it because it depends on individuals in order to exist, etc. Likewise, some things are often genuinely more absolute than others, but are not the most absolute of all; thus if we consider individuals, the species [to which they belong] is something absolute; if we consider the genus, however, it is something relative. Among measurable things, extension is something absolute, but among extended things, it is length, etc. [which is absolute]. Finally, so that it may better be understood that we are considering here a sequence of things to be known, and not the nature of each one of them, we have purposely listed *cause* and *equal* among the absolutes, even though their nature is really relative; for among philosophers, cause and effect are correlative. But here, if we ask what an effect is, one must first know the cause and not vice versa. Likewise things that are equal are correlative to each other, but we know which things are unequal only by comparing them with equals, and not vice versa, etc.

Secondly, it should be noted that there are very few pure and simple natures that can be intuited in themselves in the first place and independently of any others, either through experiences or by means of a certain light that is innate in us. But we say that these should be observed diligently; for they are the same as those that are said to be the most simple in any sequence. None of the others, however, can be perceived unless they are deduced from the most simple ones, either immediately and proximately, or only through two, three or more other conclusions. The number of these steps should also be noted, so that we know whether there are few or many steps between them and the first and most simple proposition. Such is the chain of inferences, in all cases, from which result the sequences of things to be investigated, and every question must be reduced to this if it is to be examined by a method that is certain. But since it is not easy to review all of them and, moreover, since they are not so much to be remembered as to be distinguished by a kind of sharpness of the intelligence, something should be sought which would so form the intelligence that it would recognize them whenever the need arises. I have found nothing more appropriate for this purpose, than to accustom ourselves to reflecting with some sagacity on the smallest things from among those that we have already perceived.

Finally, the third thing to notice is that one should not begin one's studies by investigating difficult things. But before committing ourselves to the determination of some question, we should first spontaneously collect, randomly, some evident truths and see if we can then deduce any other truths from them, and others again from those, and so on. Once that is done, one should reflect carefully on the truths already discovered, and think carefully about why we were able to discover some of them sooner and more easily than others, and about which ones they are, so that thereby we can judge, when we embark on some determinate questions, which of them we would be better off discovering first. For example, if the thought occurred to me that the number 6 is twice 3, I would then seek the double of 6, viz. 12; I would then look for the double of 12, if I wished, viz. 24; and the double of this, viz. 48. And thus I would deduce – something that is easy to do – that there is the same ratio between 3 and 6 as between 6 and 12, and the same ratio again between 12 and 24, etc., and therefore that the numbers 3, 6, 12, 24, 48, etc., are continuously proportional.[7] Although all of this is so clear that it seems almost childish, it follows that by reflecting carefully on it I understand how all questions that can be

RULES FOR GUIDING ONE'S INTELLIGENCE

raised about the proportions or modes of things are complex, and I understand the order in which they should be examined. This alone comprises the sum of the whole science of pure mathematics.

For I notice, first, that it was just as easy to find the double of 6 as the double of 3; likewise, in all cases, once we have found a ratio between any two magnitudes, innumerable others can be given which have the same ratio between them. The nature of the problem remains the same if three, four or more of the same kind are sought, because each one has to be discovered separately and not as a result of the others. Once the magnitudes 3 and 6 are given, I notice secondly how I would easily discover the third magnitude in a continuous proportion, viz. 12; but if the two extremes were given, namely 3 and 12, I would not as easily discover the intermediate magnitude, viz. 6. It is clear to anyone looking for the reason for this that it is a completely different type of problem from the previous one. For, in order to discover the intermediate proportional, one has to pay attention simultaneously to the two extremes and to the ratio between them, so that some new one can be calculated by division.[8] I go further and examine whether, given the magnitudes 3 and 24, I could have found as easily one of the intermediate proportionals, viz. 6 or 12. Here we encounter another kind of difficulty that is more complicated than the previous ones, namely, that one has to pay attention simultaneously, not to one or two, but to three different things in order to find a fourth. One could go even further and ask whether, given only 3 and 48, it would be more difficult still to find one of the three intermediate proportionals (viz. 6, 12 or 24). It seems initially to be more difficult. But it becomes immediately clear that this difficulty can be divided and diminished, first of all by looking for only one of the intermediate proportionals between 3 and 48, viz. 12; and by then looking for the other intermediate proportionals between 3 and 12 (viz. 6) and between 12 and 48 (viz. 24). In this way it can be reduced to the second type of difficulty mentioned above.

From all these things I realize how it is possible to follow different paths in our search for knowledge of the same thing, and that one path may be more difficult or more obscure than another. If in searching for the four continuous proportionals, 3, 6, 12, 24, we are given any two of them which are consecutive – such as 3 and 6, 6 and 12, or 12 and 24 –

it would be a very easy task to find the others from these. In that case we would say that the proportion to be discovered is investigated directly. But if two different ones are given, namely, 3 and 12 or 6 and 24, and the others have to be discovered from those, then we would say that the problem is to be investigated in the first indirect way. Likewise if two extremes were given, namely, 3 and 24, so that the intermediate proportionals 6 and 12 are sought from them, then they will be investigated in the second indirect way. Thus I could go even further and deduce many other things from this one example. But these are enough for the reader to understand what I mean when I claim that some proposition is directly or indirectly deduced, and to believe that, from very easy and primary things which are known, it is possible to discover many things even in other disciplines by careful reflections and wise distinctions.

RULE SEVEN

The completion of a science requires that all the things that are relevant to our project be reviewed, one by one, in a continuous and uninterrupted movement of thought, and that they be included in an adequate and well-ordered enumeration.

What is proposed here must be observed if we are to admit, among the truths that are certain, those that we said above cannot be deduced immediately from self-evident, first principles. This sometimes involves such a long chain of inferences that, when we arrive at the conclusion, we do not easily recall the whole journey that led us there. For that reason, we say that the weakness of our memory should be assisted by some kind of continuous movement of thought. Therefore if, for example, I first discover by various operations what is the relation between the magnitudes A and B, then between B and C, then between C and D, and finally between D and E, I do not thereby see the relation between A and E, nor can I understand it precisely from the relations that are already known unless I remember all of them. Thus I review them a number of times by a kind of continuous movement of thought,[9] simultaneously intuiting individual relations and moving on to others until I have learned to move from the first to the last so quickly that, almost without any assistance from memory, I seem to intuit the whole sequence at once. In this way, by assisting memory, the sluggishness of our intelligence is improved and its capacity is in some sense enlarged.

We add, however, that this movement should not be interrupted at any stage. For it often happens that those who try to deduce something too quickly, from remote principles, do not go through the whole chain of intermediate conclusions carefully enough to avoid skipping many of them unintentionally. It is certain, however, even when something very small is omitted, that the chain is immediately broken and the certainty of the conclusion is completely lost.

Moreover, we are claiming here that an enumeration is required for the completion of a science because, although alternative rules help to resolve other questions, it is only by using an enumeration that we can always make a true and certain judgement, whatever we apply our mind to; thus nothing will escape us, and we will seem to know something about everything.

This enumeration or induction, therefore, is such a careful and accurate inquiry into all the things that are relevant to a proposed question that we can conclude with certainty and clarity that we have omitted nothing inadvertently. Thus, each time we use it, if what we are seeking eludes us, we shall at least be wiser in knowing for certain that we could not have discovered it by any means known to us. And if by chance, as often happens, we have been able to review all the ways that are available to human beings in order to reach it, we can say with confidence that knowledge of the thing in question is completely beyond the scope of every human intelligence.

Moreover, it should be noted that by an adequate enumeration or induction we understand simply an enumeration from which the truth can be derived more certainly than from any other type of proof, apart from simple intuition. Whenever knowledge of something cannot be reduced to simple intuition, once we have rejected all the restrictions of syllogisms, this is the only path available to us to which we should entrust our full confidence. For whenever we have deduced something immediately from something else, if the inference was evident, it has already been reduced to a true intuition. If, however, we infer something from many and disparate propositions, the capacity of our intellect is often insufficient to include all of them in a single intuition. In that case, the certainty of this operation must suffice. In the same way, we cannot distinguish all the links of a rather long chain in a single glance; nevertheless, if we have seen the connection between each link and the one next to it, that will be enough for us to say that we have also seen how the last link is connected with the first one.

I have said that this operation should be adequate, because it may often be defective and consequently liable to error. For sometimes, although we have surveyed in an enumeration many things which are very evident, if we omit even the least thing the chain is broken and the certainty of the conclusion is completely lost. At other times, even if we include everything in an enumerataion that is certain but fail to distinguish the individual items from each other, we know them all only in a confused manner.

Furthermore, while this enumeration should sometimes be complete and sometimes distinct, occasionally neither is required; that is why it was only stipulated above that it should be adequate. For if I wish to prove by means of an enumeration how many kinds of entity are physical, or fall under the senses in some way, I would not claim that it was a specific number and no more, unless I first knew that I had included all of them in an enumeration and had distinguished them individually from each other. If, however, I wished to show in the same way that the rational soul is not physical, it would not be necessary for the enumeration to be complete; it would be enough to include all physical things together in various groups, and show that the rational soul could not be included in any one of them. Finally, if I wished to show by means of an enumeration that the area of a circle is greater than the area of any other figure with a perimeter of equal length, it is not necessary to list all the figures; it is enough to demonstrate this about some figures in particular, so that the same conclusion follows by induction for all other cases too.

I also added that an enumeration should be well ordered, both because there is no better remedy for the defects just listed than to examine everything in order; and also because it often happens that if particular items which are relevant to the question at issue are reviewed separately, no individual's life would be long enough, either because there are far too many items or because the same things would recur very often for examination. But if we arrange all of them in the best order, so that they are reduced in some way to various groups, it would be enough to examine one of these groups exactly, or one thing from each group, or some groups rather than others, or at least we would never examine anything needlessly a second time. This is so helpful that, as a result of good ordering, we often complete many tasks quickly and easily which on first sight seemed immense.

However, the order of things to be enumerated here can usually be varied, and it depends on each person's choice. Therefore, one should remember everything that was said in the fifth proposition[10] if an enumeration is to be done intelligently. There are many things among the more trivial of human skills such that the method of discovering them consists simply in arranging them in this order. Thus if you wish to construct the best anagram by transposing the letters of a given word, there is no need to move from the more simple to the more difficult, nor to distinguish the absolute from the relative, for these are irrelevant in this case. It would be enough to choose an order for examining the transposed letters so that the same one is never reviewed twice and the number of combinations, for example, is so distributed in different groups that it would be immediately apparent where one is most likely to find what is being sought. In this way, it will often not be a lengthy task, but merely a childish one.

Finally, these last three propositions should not be separated, because one usually has to reflect on them all together, and they all contribute equally to the perfection of the method. It would not matter which of them is taught first. We have explained them here briefly, because we have almost nothing else to do in the remainder of this treatise, where we shall explain in detail what we have dealt with in general here.

RULE EIGHT

If in a sequence of things to be investigated there is something that our intellect cannot intuit well enough, we should pause there and not examine the others that follow but, instead, we should refrain from doing fruitless work.

RULE EIGHT

The three previous rules prescribe an order and explain it; this one shows when it is absolutely necessary and when it is merely useful. For it is necessary to examine whatever constitutes an integral step in the sequence through which we have to move from something relative to something absolute (or in the opposite direction), before looking at everything else that follows. If, however, as often happens, many things belong to the same step, it is always useful to review all of them in order. But we do not have to observe this order strictly and rigidly, and frequently it is permissible to move ahead even though we know only a few or even one of them, or do not know all of them clearly.

This rule follows necessarily from the reasons given for the second rule. Nor should it be thought that this rule contains nothing new for the advancement of learning, even if it seems not to reveal any truth, but merely prevents us from investigating certain things. It does indeed teach beginners nothing else except that they should not waste their efforts – and for almost the same reason as Rule Two. But for those who know the seven preceding rules perfectly, it shows them how they can so satisfy themselves in any science that they would desire nothing further. For if someone who observes the previous rules exactly is looking for the solution of any problem, and if they are ordered to stop somewhere by this rule, they will know for certain that the knowledge they seek cannot be discovered by further labours and that this results, not from a defect of their intelligence, but because the nature of the problem itself or the human condition blocks them. This knowledge is no less scientific than that which reveals the nature of the thing itself; and anyone who would try to extend the inquiry further would seem not to be of sound mind.

All these things should be illustrated by one or two examples. If, for example, someone whose studies were limited to mathematics is looking for the line that is called the 'anaclastic' in dioptrics – the line at which parallel rays are refracted so that all of them, after refraction, intersect at a single point – they will realize easily that, according to Rules Five and Six, the determination of this line depends on the ratio between the angles of incidence and the angles of refraction. But since they will not be able to pursue this further because it belongs to physics rather than to mathesis, they will be forced to stop at the threshold; and they will make no progress if they attempt to get this knowledge by listening to philosophers or by borrowing it from experience, for that would be to violate Rule Three. Moreover, this proposition[11] is already composed and relative, and it will be explained later that it is possible to have an experience which is certain only in the case of simple and absolute things. It would also be useless to assume some ratio between the angles in question, one that one might suspect is most likely true; in that case one would no longer be searching for the anaclastic, but merely for the line that results from the logic of one's assumption.

However, for someone who has studied not only mathematics, if they wish to search for the truth about everything they encounter, in accordance with the first rule, and if they are faced with the same problem, they will also find that the ratio between the angles of incidence and refraction depends on a change in these angles that results from variations in the media; that this change in turn depends on the way in which a ray [of light] penetrates through the whole transparent body; and that knowledge of this penetration presupposes that the nature of illumination[12] is also known. Finally, in order to understand illumination, one must know what is generally meant by a 'natural power', because this is the most absolute item in the whole sequence. Therefore, once they have clearly understood what a natural power is by means of a mental intuition, they will retrace their steps in accordance with Rule Five. And if, at the second step, they are not able to discover immediately the nature of illumination, they will enumerate (in accordance with Rule Seven) all the other natural powers so as to understand this one also from knowledge of some others, at least by analogy (more about this later). Once this is done, they will seek the reason why a ray penetrates through a whole transparent medium. Thus they will follow up the remaining steps, in order, until they arrive at the anaclastic itself. Although many people have previously looked for this, I do not see that anyone who uses our method correctly can be prevented from achieving a clear knowledge of it.

But let us introduce the best example of all. If someone chooses the problem of examining all the truths for knowing which human reason is adequate (this, it seems to me, should be done once in a lifetime by all those who try seriously to acquire common sense[13]), they will find immediately from the rules given above that nothing can be known prior to the intellect, since knowledge of everything else depends on it and not vice versa. Thus having reviewed all the things that follow directly from knowledge of the pure intellect, they will enumerate among the rest whatever other instruments for knowing we possess apart from the intellect; there are only two of them, namely, phantasy and sensation. They will therefore apply all their efforts to distinguishing and examining those three ways of knowing; once they recognize that truth or falsehood in a strict sense can be present only in the intellect, but that they often have their roots in the other two, they will pay careful attention to avoid anything by which they can be deceived. They will enumerate precisely all the ways in which human beings can reach the truth, so that they can follow one which is certain. There are not so many of these that it would be impossible to discover all of them by means of an adequate enumeration. This will seem extraordinary and incredible to those who are inexpert. As soon as they have distinguished the knowledge of individual objects which merely fills and adorns the memory from that which really entitles someone to be called more learned – something they can do easily [. . .][14] they will feel that they are no longer ignorant of anything because of a defect of intelligence or art, and that henceforth there is nothing that anyone could know which they would not be capable of knowing too, on condition merely – as is appropriate – that they apply their mind to it. There may often be many problems proposed that they are prohibited from resolving by this rule; however, since they will perceive clearly that such problems go beyond the scope of every human intelligence, they will not think of themselves as more ignorant than others. The very fact that they know that the thing being sought cannot be known by anyone will be abundantly adequate to satisfy their curiosity once it is reasonable.

To avoid being always uncertain about what the mind can do and working vainly and with temerity, before applying ourselves to know things in particular, we ought, once in a lifetime, to inquire carefully what kind of knowledge human reason is capable of. To do this better, one should always seek first whatever is more useful among those things that are equally easy.

This method is similar to those mechanical arts that do not need any other arts, but rely on their own resources for manufacturing their instruments. If someone wished to practise one of these – for example, to practise the art of a blacksmith – and if they had none of the tools, they would be forced initially to use a hard stone or some rough lump of iron as an anvil, to use a rock in place of a hammer, to adapt wood as tongs and to collect other similar things as the need arises. Once these are all ready, they would not immediately try to forge swords, helmets or other things made of iron to be used by other people. Before all else they would make a hammer, anvil, tongs and the other things that they need themselves. We learn from this example that since we have been able to find, in these preliminary discussions, only certain hidden rules that seem to be innate in our minds rather than prepared by using some art, we should not immediately use them to try to adjudicate the disputes of philosophers or to resolve the knots of mathematicians; they should be used instead to seek out others, with the greatest care, which are more necessary for the examination of the truth, especially since there is no reason why it would seem to be more difficult to discover them than any of the questions that are usually proposed in geometry, physics or other disciplines.

But there is nothing that one can more usefully ask, in this context, than: what is human knowledge, and how far does it extend? We are including both in a single question here, and we think that it is the first question to be investigated by means of the rules already provided. It is also something that should be done once in a lifetime by anyone who has the slightest love of truth, because the true instruments for learning and the whole method are contained in such an investigation. Nothing seems to me to be more inept, however, than to dispute confidently about nature's secrets, the influence of the skies over the lower regions, the prediction of future things and so on, as many people do, and not to have ever asked whether human reason is adequate to discover such things. Nor should it be seen as a difficult or arduous task to define the limits of an intelligence that we experience in ourselves, since we frequently do not hesitate to make judgements about things that are external and very foreign to us. It is not an immense task either, to want to comprehend in our thought all the things that are contained in the universe, so as to know only how individual things may be subject to examination by our mind. For nothing can be so complex or dispersed that it cannot be circumscribed by the enumeration that we discussed, and arranged under a certain number of headings. To test this in relation to the question raised, let us first divide everything that is relevant to it into two parts. Everything should be referred either to us, who are capable of knowing, or to the things themselves that can be known. We shall discuss these two parts separately.

Now we recognize in ourselves that the intellect alone is capable of scientific knowledge, but it can be aided or hindered by three other faculties, namely, by imagination, sensation and memory. One must therefore consider, in order, how each one of these faculties can impede us, so that we can take precautions, or how they can help us, so that we can exploit all their resources. But this part will be discussed by means of an adequate enumeration, as the following proposition shows.[15]

One must then look at the things themselves, and consider them only in so far as they can be reached by the intellect. From this perspective, we divide them into the most simple natures, and into complex or composite natures. Among simple natures there can only be spiritual or physical, or those that belong to both. Then, among composite natures, the intellect experiences some things as composite before deciding to determine anything about them; in the case of others, the intellect itself composes them. All these things will be explained more fully in the twelfth proposition [i.e. Rule Twelve], where it will be demonstrated that falsehood can occur only in those that are composed by the intellect. We therefore distinguish composite natures in turn into those which are deduced from very simple and self-evident natures – we shall discuss them in the whole following book – and those which presuppose other natures, and which we experience as composite in the things themselves – and we shall devote the whole third book to explaining them.[16]

In this whole treatise, we shall try to follow closely and to display as easy all the paths to knowledge of the truth that are available to human beings, so that whoever learns this whole method perfectly, no matter how mediocre their intelligence may be, will see that there are no paths that are more closed to them than to other people, and that there is nothing else of which they are ignorant due to a defect of intelligence or art. Instead, whenever they apply their mind to knowing something, they will either discover it immediately, or they will see with certainty that it depends on some experience that is not within their power, and therefore they will not blame their own intelligence, even if they are forced to stop at that point; or, finally, they will demonstrate that the thing they are seeking exceeds the scope of every human intelligence and therefore they will not think of themselves as being more ignorant, because there is as much scientific knowledge in this realization as in anything else they may have known.

RULE NINE

One should turn the full insight of one's intelligence to the smallest and simplest things, and dwell on them long enough to become used to intuiting the truth in them distinctly and clearly.

Having explained the two operations of our intellect, intuition and deduction, that we said were the only ones to be used for acquiring scientific knowledge, we proceed in this and the following proposition to explain how, by practising, we can improve at using these operations and, at the same time, how we can cultivate the two principal faculties of our intelligence, namely, clarity in intuiting particular things and wisdom in deducing some things skilfully from others.

Indeed, we learn how mental intuition should be used by comparing it with ocular vision. For whoever tries to look at many objects simultaneously in a single glance sees none of them distinctly. In the same way, anyone who usually attends to many things in a single act of thought confuses their intelligence. But artisans who are involved in detailed work and are used to directing their keen vision attentively to individual points acquire, by practice, the ability to distinguish things perfectly no matter how fine and delicate they are. Similarly, those who never distribute their thought between many objects at the same time, but always focus it completely on thinking about certain very simple and easy things, become insightful.

RULE NINE

It is a common mistake among mortals that more difficult things seem more attractive to them. Most people think that they know nothing when they see a very clear and simple cause of something; but, at the same time, they admire very abstract and far-fetched arguments of philosophers, even though they are usually based on foundations which have never been adequately understood by anyone; they are hardly sane who think that shadows are clearer than the light. It should be noted, however, that those who have genuine knowledge can discern the truth equally easily, whether they get it from a simple object or from an obscure one. For they comprehend every truth by means of a similar, unique and distinct act once they have found it. The difference lies entirely in the path used – and this should surely be longer, if it leads from the first and most absolute principles to a more remote truth.

Everyone, therefore, should become accustomed to comprehending in their thought things which are simultaneously so simple and so few that they never think they know anything, unless it is intuited as distinctly as what they know most distinctly of all. Some people are indeed born much better at this than others, but one's intelligence can become much better at it by practice and by art; and there is one thing which, it seems to me, is most commendable of all, namely, that everyone should become convinced that the sciences, no matter how hidden they may be, are to be deduced not from lofty and obscure things, but merely from those that are easy and most evident.

For example, if I wished to investigate whether some natural power could travel instantaneously to a distant place right through a particular medium, I would not turn my mind immediately to the power of a magnet or the influence of the stars, nor even to the speed of illumination, to find out if actions like these happen to be realized in an instant. For it would be more difficult to prove this than what I am looking for. I would reflect instead on the local motions of bodies, because there could be nothing easier to perceive in this whole area. And I shall notice that a stone cannot really move from one place to another in an instant, because it is a body; a power, however, which is similar to that which moves the stone, is communicated in only an instant if the power alone is to move from one body to another.[17] For example, if I move one end of a stick, no matter how long it is, I easily understand that the power by which that part of the stick is moved necessarily moves all the other parts of the stick in one and the same instant, because in that case the power is communicated on its own and it does not exist in any other body, such as a stone, by which it is carried.

In the same way, if I wish to know how opposite effects can be simultaneously produced by one and the same simple cause, I shall not borrow from physicians' drugs which expel certain humours while retaining others; nor shall I talk at length about the moon, that it warms things by its light while it cools them by means of some occult quality. Rather, I shall look carefully at a balance on which the same weight, at one and the same instant, elevates one side while it depresses the other, and similar cases.

RULE TEN

In order to sharpen the intelligence, it should be exercised in searching for things that have already been discovered by others, and it should review methodically even the most trivial results of human skill, especially those that deploy or presuppose order.

I confess that I was born with an intelligence such that my greatest pleasure in studying has always been, not in hearing the reasoning of others, but in discovering the same reasoning by my own efforts. This alone attracted me to study while I was still young; thus whenever the title of some book promised a new discovery, I would try before reading any further to see if I could possibly discover something similar by a kind of innate sagacity, and I took special care that a hasty reading would not deprive me of such an innocent pleasure. That was successful so often that I eventually noticed that I was no longer reaching the truth of things, as others usually did, by means of wandering, blind inquiries that are assisted by luck rather than by art, but I discovered through long experience certain rules that were very useful, which I subsequently used in thinking about others. I thus carefully developed this whole method, and became convinced that the method of study that I followed from the beginning was the most useful of all.

RULE TEN

But since every intelligence is not as disposed by nature to investigate things by its own efforts, this proposition teaches that we should not initially undertake the most difficult and arduous things, but should first tackle some very simple and trivial arts, and primarily those in which order is most evident – such as those of artisans who weave tapestries and carpets, or of women who do needlepoint or weave threads of various textures in infinitely many ways; similarly, all games of number and whatever pertains to arithmetic, and the like. It is surprising how much all these things exercise one's intelligence, on condition that we do not borrow their discovery from others but from ourselves. Since there is nothing in these that remains hidden, and they are completely adapted to the capacity of human knowledge, they distinctly reveal innumerable examples of order to us, all different from each other but, despite that, all rule-governed, in the proper observance of which almost the whole of human sagacity consists.

This is why we recommended that things should be investigated methodically, and this is usually nothing more, in the case of trivial things, than the constant observance of order – an order that exists in the thing itself, or is subtly introduced into it. For example, if we wish to read some writing that has been camouflaged by unknown characters, there is evidently no apparent order in it, but we should invent one, both to test every assumption that can be made about individual characters, words or sentences, and also to arrange them in such a way that we could know by means of an enumeration what can be deduced from them. Above all we should avoid, in cases like this, wasting our time by guessing randomly and without any art; for even though they can often be discovered without art and, for those who are lucky, they are sometimes solved more quickly than by using a method, they obscure the light of intelligence and make it so accustomed to childish and vain things that subsequently it always stays on the surface of things and cannot penetrate any further into them. At the same time we should not fall into the mistake of those who occupy their thought exclusively with serious and deeper things of which, despite looking for a profound scientific knowledge, they acquire only a confused knowledge after much labour. Therefore we should practise first with easier things, but we must do so methodically so that we get used to penetrating always to the inner truth of things, as if we were playing games. In this way we shall gradually find later – and in a shorter time than we could hope for – that we can also deduce many propositions which seem to be very difficult and complicated from clear principles.

However, some people may be surprised that here, where we are looking for a way to make ourselves more capable of deducing some truth from others, we omit all the rules of dialecticians[18] with which they think they govern human reason by prescribing forms of discourse that conclude with such necessity that, if reason were to rely on them, then, even if it were distracted in some way from considering an inference attentively and clearly, it could still reach a conclusion with certainty merely in virtue of the form. But we have recognized that truth often escapes from these bonds, while those who use them are meanwhile left entrapped by them. This does not happen as frequently to others, and we find by experience that the most acute sophisms usually entrap almost no one who uses pure reason, but only the sophists themselves.

That is why we reject those forms as obstacles to our project, primarily by guarding against our reason being distracted while we are investigating the truth about something. We look instead for all the aids by which our thought may be held attentively, as will be shown in what follows. But to make it even more evident that this art of discourse contributes nothing at all to knowledge of the truth, it should be realized that, by using their art, dialecticians cannot construct any syllogism that concludes in the truth unless they already had its matter,[19] that is, unless they already knew the very truth that is deduced in the syllogism. It is clear from this that they can learn nothing new from such a form, and therefore that the common dialectic is completely useless for someone who wishes to investigate the truth of things, but that it can be used merely to expound, to others, reasons that are already known. It should therefore be transferred from philosophy to rhetoric.

RULE ELEVEN

Once we have intuited some simple propositions, if we conclude something else from them, it is useful to review them in a continuous and uninterrupted movement of thought in order to reflect on their mutual relations and, in so far as this is possible, to conceive distinctly of many of them simultaneously; for that is how our knowledge becomes much more certain, and the capacity of our intelligence is greatly increased.

This is the time to explain more clearly what was said earlier about mental intuition in Rules Three and Seven. For in one place we contrasted intuition with deduction and, in another, merely with enumeration (which we defined as an inference from many and disparate things). But we said in the same place that a deduction of one thing from another is accomplished by means of an intuition.

It had to be done in that way because we require two things for a mental intuition, namely, that one understands a proposition clearly and distinctly and, secondly, that it be understood in its entirety all at once and not piecemeal. But if we think of making a deduction, as in Rule Three, it does not seem to be done all at once; instead, it involves some kind of motion of our intelligence which infers one thing from another, and therefore we correctly distinguished it from intuition. If, however, we consider a deduction as already completed, as we did in Rule Seven, then it no longer implies a movement, and therefore we suppose that it is seen by means of an intuition when it is simple and clear, but not when it is complex and convoluted. We have given the name 'enumeration' or 'induction' to the latter, because in that case the whole deduction cannot be comprehended simultaneously by the intellect, but its certainty depends in some way on memory, in which judgements about the individual parts of the enumeration must be stored so that, from them, a single thing is derived from all of them together.

All these had to be distinguished to interpret this rule. For since the ninth [rule] dealt with intuition alone, and the tenth [rule] was concerned only with enumeration, this rule explains how these two operations mutually assist and support each other, so that they seem to merge into one by a certain movement of thought which attends carefully to individual things and simultaneously moves on to others.

We find that it is useful in two ways, namely, for knowing the conclusion that we are concerned about, and for making the intelligence more fit to discover other things. For when memory is weak and unstable – and we said that the certainty of conclusions which include more than we can grasp in a single intuition depends on memory – it should be refreshed and strengthened by this repeated and continuous movement of thought, so that if I discover through many operations what is the relation between the first and second magnitude, then between the second and third, then between the third and fourth, and finally between the fourth and fifth, I do not thereby see what is the relation between the first and the fifth, nor can I deduce it from what is already known about the other relations unless I recall all of them. Therefore I have to review them by a repeated thought until I can move from the first to the last so quickly that, leaving almost no parts in my memory, I seem to intuit the whole thing at once.

Everyone sees how the sluggishness of one's intelligence is thus corrected, and its capacity is also increased. But it should also be noted that the greatest benefit of this rule consists in the fact that, by reflecting on the mutual dependence of simple propositions on each other, we acquire the habit of distinguishing quickly what is more or less relative and by what steps they are reduced to what is absolute. For example, if I review some magnitudes that are continuously proportional, I shall reflect on all the following: by a comparable conception, which is neither more nor less easy, I recognize the relation between the first and the second, the second and the third, the third and the fourth, and so on; I cannot as easily conceive of the relation between the second and the first and third at the same time, and it is much more difficult again to conceive of the relation of the same second magnitude to the first and fourth, etc. From these, therefore, I know why, if only the first and second are given, I can easily find the third and fourth, etc. – namely, because this is realized by individual and distinct conceptions. But if the first and third alone are given, it is not so easy to discover the intermediate magnitude, because this can be done only by means of a conception that simultaneously involves two of the previous conceptions. And if only the first and fourth are given, it is even more difficult to intuit the two intermediate magnitudes, because here three simultaneous conceptions are involved. Thus it follows that it would seem to be more difficult to discover three intermediate magnitudes between the first and the fifth; however, there is another reason why this happens to be otherwise, i.e. that although there are four conceptions joined together here simultaneously, they can nevertheless be separated, since four is divisible by another number. Thus I could simply look for the third magnitude from the first and the fifth, and then the second one from the first and the third, and so on. Anyone who has become used to reflecting on these and similar things will recognize immediately, when they investigate a new question, the source of the difficulty and the simplest way for resolving it. This is the greatest aid to knowledge of the truth.

RULE TWELVE

Finally, one should use all the assistance of the intellect, imagination, sensation and memory, to intuit simple propositions distinctly; to discover what one is searching for by comparing it with what is already known; and to find which things should be combined with each other, so that nothing is omitted from human diligence.

This rule concludes everything that was said above, and teaches in a general way what needs to be explained in detail in what follows.

Only two things are relevant for knowledge of things, namely, we who know, and the things themselves to be known. We have only four faculties in us that we can use for this purpose, namely, the intellect, imagination, sensation and memory. The intellect alone is capable of perceiving the truth, but it must be assisted by the imagination, sensation and memory, so that we do not happen to omit anything that was provided among our powers. From the point of view of the things [to be known], it is enough to investigate three factors: viz. first, that which is evident in itself; then, how one thing can be known from something else; and what can be concluded from both of them. This enumeration seems to me to be complete, and not to omit anything to which human diligence can be applied.

RULES FOR GUIDING ONE'S INTELLIGENCE

Turning then to the first, I would prefer to explain here what the human mind is, what the body is, and how the latter is informed by the former;[20] what are the faculties in the whole composite [of body and mind] that are used to know things; and what each faculty does, if I had not thought I would not have enough space in which to include all the things that must be presupposed in order to make the truth about all these issues clear to everyone. For I wish to write always so as not to make claims about anything that can lead to controversy, unless I first set out the reasons which have led me to it and by which I think others can also be convinced.

But since that is impossible here, it will be enough for me to explain as briefly as I can what is the most useful way of conceiving everything in us that is used for knowing things. Do not believe that things are as I describe them unless you want to; but what would prevent you from following these assumptions if it seems that they take nothing away from the truth of things but rather make everything clearer? This is no different than if you were to assume something about a quantity in geometry by which the force of a demonstration is not in any way diminished, even though you often think otherwise about the same nature in physics.

First, then, one must understand that all the external senses, in so far as they are parts of the body – although we apply them to objects through an action, viz. through local motion – have sensory perceptions in the strict sense by means of a passion, in the same way as wax receives an impression from a seal. One should not think that I am speaking by analogy here; clearly, it should be understood that the external shape of the sentient body is really changed by the object, in exactly the same way as that which is on the surface of the wax is changed by the seal. This must be accepted not only when we touch some body which has a shape, or is hard or rough, etc., but even when, as we touch something, we perceive heat or cold, and similar things. The same applies to other senses; the first opaque membrane in the eye receives the shape impressed on it in this way by an illumination endowed with various colours, and the first membrane in the ear, nose and tongue which is impervious to an object likewise assumes a new shape from a sound, odour and flavour.

It helps a great deal to conceive of all these things in this way, because nothing is more accessible to the senses than shape, for it is both touched and seen. And from the fact that the concept of shape is so common and simple that it is involved in everything that is capable of being sensed, it is possible to demonstrate that this assumption has no more false consequences than any other. For example, you may imagine colour to be anything you wish, but you will not deny that it is extended and that it therefore has a shape. As long as we take care not to invent any new entity uselessly and foolishly, while not denying anything that others have preferred to claim about colour, what difficulty could result if we merely abstract from everything else apart from the fact that colour has the nature of shape and if we conceive the difference between white, blue, red, etc., as being like that between the following shapes, or similar ones:

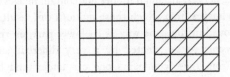

The same thing can be said about all of them, since it is certain that the infinite number of shapes suffices to express all the differences between things that can be sensed.

Secondly, one should conceive that, when an external sense is moved by an object, the shape that it receives is passed on to some other part of the body that is called the common sense;[21] this occurs at the same instant and without any real entity moving from one to the other, in exactly the same way that now, as I am writing, I understand that at the same moment that the letters are expressed on the paper, not only is the lower part of the pen moved, but it is impossible for the slightest movement to occur there without being received simultaneously in the whole pen. All the diversity of those motions is traced out in the air by the upper part of the pen, even though I do not conceive of anything real passing from one end of the pen to the other. However, is there anyone who thinks that there is less connection between the parts of the human body than between those of a pen, and who can think of some simpler way of expressing it?

Thirdly, the common sense should be thought of as functioning like a seal to form, in the phantasy[22] or imagination as if in wax, all those shapes or ideas that arrive, pure and without a body, from the external senses. This phantasy is a genuine part of the body, and it is big enough for various parts of it to assume many shapes that are distinct from each other, and to retain them usually for some time; in that case, it is the same as what is called 'memory'.

Fourthly, it should be conceived that the motive force or the nerves themselves originate from the brain, where the phantasy is, and they are moved by it in various ways, as the external senses move the common sense or the whole pen is moved by its lower part. This example shows once again how the phantasy can be the cause of many motions in the nerves, although it does not contain images of them but certain other images from which those motions can result. Nor is the whole pen moved in the same way as its lower part; on the contrary, most of it seems to be affected by a very different and opposite movement. From these comments one can understand how all the motions of other animals can occur, even granting that there is absolutely no knowledge of things in them, but only a completely corporeal phantasy. It also allows us to understand how all the operations that we perform without any assistance from reason occur in ourselves.

Fifthly, and finally, we should conceive that the power by which we know things in a strict sense is completely spiritual, and that it is as distinct from the whole body as blood is from bone, or a hand from an eye; that it is one and the same power which, together with the phantasy, receives shapes[23] from the common sense or applies itself to those that are stored in memory, or forms new shapes which so occupy the imagination that it is often no longer capable of receiving ideas coming from the common sense, or of transmitting them to the force responsible for motion in the way in which purely bodily motions take place. In all these things this cognitive power is sometimes passive, at other times active; it sometimes resembles a seal, at other times wax. However, this should be understood merely as an analogy here, and nothing exactly like this occurs in physical things. But it is one and the same power which, if it applies itself together with the imagination to the common sense, it is said to see, to touch, etc.; if it applies itself to the imagination alone, in so far as it is endowed with various shapes, it is said to remember; if it applies itself to the same faculty in order to construct new shapes, it is said to imagine or conceive; finally, if it acts alone, it is said to understand. How this last operation is accomplished, I shall explain in greater detail in the appropriate place. The same power therefore is called 'pure intellect', 'imagination', 'memory' or 'sensation', depending on its different operations. But it is called 'intelligence', in a strict sense, when it forms new ideas in the imagination or applies itself to those already formed there. We think it is capable of these different operations, and the distinction between these names should be observed in what follows. Once all these things have been understood in this way, the attentive reader will easily conclude what assistance could be provided by each faculty, and will know the extent to which human effort can be stretched to compensate for deficiencies of intelligence.

For just as the intellect can be moved by the imagination or, on the contrary, can act on it, so likewise the imagination can use a motive force to act on the senses by applying them to objects and, on the contrary, the senses can act on it by depicting the images of bodies on it. But memory, at least that which is physical and similar to the recall of brute animals, is not distinct from the imagination. It follows certainly that if the intellect is applied to things in which there is nothing physical or anything similar to what is physical, it cannot be assisted by those faculties. But, on the contrary, in order not to be impeded by them, the senses should be shut down and the imagination should be cleansed of every distinct impression, in so far as that is possible. If, however, the intellect plans to consider something that can be related to a body, the idea of that thing should be formed in the imagination as distinctly as possible. And to do that more easily, the thing itself that the idea represents should be presented to the senses. A plurality of things cannot assist the intellect to intuit individual things distinctly. In order [for the intellect] to deduce a single thing from many things – something that has to be done frequently – anything in the ideas of things that does not require one's attention at the time should be rejected, so that everything else is more easily retained in one's memory. In the same way, the things themselves should then not be presented to the external senses but rather some simplified representations of them. Once the latter are enough to avoid a lapse of memory, the more simplified they are the better. Anyone who observes all these things, will seem to me not to have omitted anything that is relevant to this section.

To proceed now to the second question[24] – i.e. to distinguish carefully the notions of simple things from those which are composed from them; to see where falsehood may occur in both, so as to guard against it; and to see what things can be known with certainty, so as to concentrate on those alone – there are some things that must be assumed here, as in the discussions above, which may not perhaps be accepted by everyone. But that is not important, even if they are believed to be no more true than those imaginary circles by which astronomers describe their phenomena, on condition that, by using them, you distinguish between true and false knowledge about anything.

We say first, therefore, that to view things from the perspective of our knowledge is different from speaking about them as they really are. For example, if we think of some body that is extended and has a shape, we shall admit that – from the perspective of the thing itself – it is something that is one and simple and, in this sense, it cannot be said to be composed of the natures of body, extension and shape, because they have never existed as parts that were distinct from each other. But from the perspective of our knowledge, we call such a body a composite of these three natures, because we understood each of them separately before we were able to judge that the three of them could occur together in one and the same subject. Therefore, since we are discussing things here only in so far as they are understood by the intellect, we apply the term 'simple' only to those the knowledge of which is so clear and distinct that they cannot be divided by the mind into other things that are more clearly known. Shape, extension, motion, etc., are examples of this. We conceive of all other things, however, as being in some way composed of them. This should be understood so generally that no exception is made, even for those that we sometimes abstract from the simple things themselves – for example, if we say that shape is the limit of an extended thing, thinking of 'limit' as something more general than shape, because one can also speak about the limit of a duration, of a movement, etc. In this case, although the meaning of 'limit' is abstracted from a shape, it should not therefore be thought of as simpler than shape; rather, since it is attributed to other things, such as the extremity of a duration or a movement, etc., which are things that are completely different from shape, it ought to be abstracted from those also; consequently it is something composed of many natures which are very different and to which it is applied only equivocally.

We say, secondly, that the things that are said to be simple from the perspective of our intellect are either completely intellectual, or completely material, or common. The completely intellectual are those that are known by the intellect by means of a certain innate light, without any assistance from a physical image. It is certain that there are some things like that; no physical idea can be framed which could represent to us what knowledge, doubt or ignorance are, nor what is an act of the will that could be called a volition, and such things. All of these are things that we genuinely know, and we know them so easily that, to do so, it is enough for us to be reasonable. The completely material are those that are known to be present only in bodies, such as shape, extension, movement, etc. Finally, those are said to be 'common' which can be attributed indiscriminately at different times to bodily things and to spirits, such as existence, unity, duration and similar things. We should also include here the common notions that are like certain links for joining together other simple natures, and on the evidence of which depends whatever we conclude by reasoning. For example: 'those which are the same as a third thing are the same as each other' and 'things which cannot be related in some way to a third thing are also different from each other', etc. These common simples can be known either by the intellect alone, or by the intellect intuiting images of material things.

Moreover, it is also appropriate to classify among these simple natures their privations or negations, in so far as they are understood by us; for the cognition by which I intuit what nothing is, or what an instant of rest is, is no less genuine than that by which I understand what existence, duration or a movement are. This way of understanding here will help us to say later that all the other things that we know are composed of these simple natures. Thus if I judge that some shape is not moved, I shall say that my knowledge is in some way composed of a shape and of rest; and likewise for other cases.

We say, thirdly, that all these simple natures are known in themselves, and that they never contain any falsity. That is easily shown if we distinguish the faculty of the intellect by which it intuits and knows things, from the faculty by which it judges when affirming and denying. For it can happen that we think we do not know the things that we genuinely know when, besides what we intuit or what we reach by our understanding, we suspect there is something else in them that is hidden from us, and that our knowledge may be false. For this reason it is clear that we are mistaken if we ever judge that there is something in these simple natures that cannot be completely known by us. For if we have even a slight mental grasp of it – and this must be the case since we are assuming that we make some kind of judgement about it – it follows from this alone that we know it completely. Otherwise it could not be said to be simple, but composed of what we understand about it and of what we judge we do not know about it.

We say, fourthly, that the conjunction between these simple things is either necessary or contingent. It is necessary when one is so included in the concept of the other, in some confused way, that we cannot conceive of one distinctly if we judge that it is distinct from the other. Shape is united with extension in this way, and movement with duration or time, etc., because it is impossible to conceive of a shape deprived of all extension or of a movement that has no duration. Likewise if I say that 4 and 3 are 7, this conjunction is necessary; for we do not conceive of 7 distinctly, unless we include 4 and 3 in it in a confused way. In the same way anything that is demonstrated about shapes or numbers is necessarily joined with whatever the demonstration is about. Nor is this necessity found only in things that can be perceived. For example, if Socrates says that he doubts everything, it follows necessarily that he at least understands that he doubts; likewise, he knows consequently that something can be either true or false, etc., since these are necessarily connected to the nature of doubt. However, things are united in a contingent way when they are not connected by any indissoluble relation; for example, when we say that a body is alive, that a man is dressed, etc. But there are many things that are often joined together necessarily, which are classified as being contingently joined by many people who do not notice the relation between them; for example, this proposition: 'I am, therefore God exists.' Also, 'I understand, therefore I have a mind that is distinct from the body,' etc. Finally, it should be noted that, in the case of many propositions that are necessary, their converse is contingent; for example, although I conclude with certainty that God exists from the fact that I am, it is not the case that I can affirm that I exist from the fact that God is.

We say, fifthly, that we can never understand anything apart from these simple natures, or some mixture or composition of them; and it is often easier to pay attention simultaneously to a number of them joined together than to separate one from another. For example, I can know a triangle even if I have never thought that knowledge of an angle, a line, of the number three, of shape, of extension, etc., is also included in that knowledge. But that does not prevent us from saying that the nature of a triangle is composed of all those natures, and that they are better known than the triangle, because those are the very natures which are understood in it. The same triangle may also include many other natures that are hidden from us, such as the size of the angles, which are equal to two right angles, and the innumerable relations between the sides and the angles, or how big its area is, etc.

We say, sixthly, that those natures that we call 'composite' are known by us either because we experience what they are like, or because we compose them ourselves. We experience everything that we perceive by sensation, anything we hear from others, and in general, whatever reaches our intellect either from outside itself or from its reflective contemplation of itself. It should be noted in this context that the intellect can never be deceived by any experience, on condition that the thing which is its object is intuited exactly as it holds it either in itself or in an image; that it does not judge that the imagination faithfully represents the objects of the senses or that the senses are endowed with the true shape of things; and, finally, it does not judge that external things are always as they appear. For we are liable to make mistakes in all of these. For example, if someone tells us a story and we believe that the event happened that way; or if someone who is suffering from jaundice judges that everything is yellow because their eye is tinged with yellow; or, finally, when the imagination is malfunctioning, as happens in the case of melancholics, if we judge that its disturbed images represent real things. But these same things will not deceive the intellect of someone who is wise, because they will judge that whatever they accept from their imagination is really represented there, but they will never claim that the very same thing was transmitted in its integrity and without any modification from external things to the senses and from the senses to the phantasy, unless they know this independently on the basis of some other evidence. On the other hand, we ourselves compose the things we understand, whenever we believe that they contain something that has not been perceived immediately by our mind through some experience. For example, if someone who has jaundice convinces themselves that the things they see are yellow, this thought of theirs would be composed from what their phantasy represented to them and what they assumed on their own initiative, viz. that a yellow colour appears, not because of some defect of the eye but because the things seen really are yellow. It follows therefore that we can be mistaken only when the things that we believe are in some way composed by ourselves.

We say, seventhly, that this composition can be realized in three ways, viz. by impulse, by conjecture or by deduction. People compose their judgements about things by impulse when they are moved by their intelligence to believe something without being convinced by any reason and are simply determined by some superior power,[25] by their own freedom or by the condition of their phantasy. The first one is never mistaken, the second is rarely, and the third is almost always so. However, the first of these is not relevant in this context, because it does not come within the scope of any method. Composition results from conjecture if, for example, from the fact that water is further from the earth's centre than earth and is also a more subtle substance, and also that air is higher again than water and is also more rare than it, we conjecture that there is nothing beyond the air except some very pure ether, and that it is much more subtle than the air itself, etc. What we compose by this kind of reasoning does not mislead us if we consider it merely as probable and never affirm it as true, but it does not make us any wiser.

There remains only deduction by which we can compose things in such a way that we are certain of their truth. However, there may be a number of defects in it too. For example, if from the fact that we perceive nothing by sight, by feeling or by any other sensation in a space that is full of air, we conclude that it is empty by joining together mistakenly the natures of a vacuum and of this space. The same occurs any time we judge that we can deduce something general and necesssary from a particular or contingent thing. But it is within our power to avoid this mistake, by never joining things together unless we see that the conjunction of one with the other is completely necessary; for example if, from the fact that shape has a necessary connection with extension, we deduce that nothing can have a shape if it is not extended, etc.

It follows from all this that, first, we have expounded distinctly and I think by means of an adequate enumeration what we had been able to show at the beginning only confusedly and approximately; that is, that there are no paths available to human beings to a certain knowledge of the truth apart from clear intuition and necessary deduction, and also what are those simple natures that were mentioned in the eighth proposition. It is clear that mental intuition extends to all those natures, to know the necessary connections between them, and also to all the things that the intellect experiences as being either in itself or in the phantasy distinctly. More will be said about deduction below.[26]

It follows, secondly, that no effort is required to know those simple natures because they are adequately known in themselves, but only to distinguish them from each other and to intuit them separately by a focused eye of the mind. For no one has such a dull intelligence that they do not perceive that, when they are sitting down, they are in some way different from when they are standing up. But not everyone distinguishes equally clearly the nature called 'position' from whatever else is included in that thought, nor can they claim that it is the position alone that is changed. This is not redundant advice here, because often those who are educated are used to being so clever that they find a way of blinding themselves even about things that are self-evident and are never unknown to the uneducated. This happens whenever they try to explain things that are known in themselves by means of something else which is more evident. Then either they explain something else, or nothing at all. For is there anyone who does not perceive everything – whatever it happens to be – that is involved in a change when we change place, and who would understand the same thing if they were told that 'place is the surface of the surrounding body'?[27] For that surface may change while I remain immobile and do not change place; on the other hand, it may be moved with me in such a way that, although the same surface surrounds me, I am no longer in the same place. But do they not seem to utter magical words with an occult power that is beyond the capacity of human intelligence when they say that 'motion' – something that everyone knows very well – 'is the act of a a being in potency in so far as it is in potency'?[28] Is there anyone who understands these words, or anyone who does not know what motion is? And who would not concede that they are inventing a problem where none exists? It should be said, therefore, that things should never be explained with such definitions, lest we apprehend composite things instead of simple natures; but we should intuit only the latter, separated from all others, attentively and by the light of our intelligence.

It follows, thirdly, that the whole of human science consists in this one thing, that we see distinctly how those simple natures combine to compose other things. It is very important to be aware of this. For whenever some problem is proposed for examination, almost everyone stops at the threshold, uncertain about which thoughts they should provide for their mind and keen to look for some new kind of being that they previously did not know. For example, if someone asks what is the nature of a magnet, they anticipate that it will be something difficult and inaccessible, and they immediately turn their mind away from everything that is well known and turn it towards some very difficult things. By wandering about in the empty space of many causes, they hope to find by chance something novel. But someone who believes that nothing in the magnet can be known which is not constituted by various simple natures that are known in themselves is not uncertain about what should be done; rather, they first collect all the experiences that can be had about this stone, from which they then try to deduce what combination of simple natures is necessary to produce all the effects that are experienced in this stone. Once this is found, they can boldly claim that they have truly perceived the nature of a magnet in so far as it is humanly possible to discover it from the available experiences.

Finally, it follows fourthly from what has been said that knowledge of some things should not be thought to be more obscure than knowledge of others, since all kinds of knowledge are of the same nature and consist merely in composing things that are known in themselves. Hardly anyone acknowledges this, but convinced in advance of the opposite view, those who are more shameless take the liberty of claiming that their conjectures are genuine demonstrations and, even in the case of things that they know nothing about, they often claim to see obscure truths as if through a cloud. And they are not bashful in proposing these, linking their concepts with certain words by the aid of which they are accustomed to discuss many things and to speak logically. But neither they themselves nor their hearers understand them. Those who are more modest, however, often refrain from examining many things – even if they are not difficult and are very necessary for life – simply because they think they are not competent to do so; and because they think that such issues can be understood by others who are endowed with more native intelligence, they adopt the views of those in whose authority they have most confidence.

We say, eighthly,[29] that the only possible deductions are as follows: from words to a thing; from an effect to a cause, or from a cause to an effect; from like to like; or from parts to parts, or to the whole itself [...][30]

For the rest, lest the connections between our rules be hidden from anyone, we divide everything that can be known into simple propositions and questions. As regards simple propositions, the only rules we offer are those that prepare the faculty of knowing to see more distinctly and examine more skilfully any object one wishes, because these propositions should occur to us spontaneously and cannot be sought. We have dealt with that in the first twelve rules, and we think we have provided everything that we think is capable of making the use of our reason in any way easier. As regards questions, some may be understood perfectly, even if we do not know their solutions, and we discuss those alone in the next twelve rules. There are others, finally, that are not perfectly understood, and we defer them to the final twelve rules. We have invented this division purposely so that we are not forced to say anything that presupposes knowledge of what follows, and we may teach first of all those things that we think we should initially address in order to cultivate intelligence. It should be noted that, among the questions that are perfectly understood, we classify only those in which we perceive three things distinctly: viz. what are the signs by which the thing being sought can be recognized; what precisely is it from which we should deduce it; and how can it be proved that one depends on the other in such a way that one of them cannot be in any way changed if the other remains unchanged. Thus we have all the premisses, and the only thing that remains to teach is how the conclusion should be found – not that one thing is to be deduced from one simple thing (it has already been said that this can be done without rules), but by unravelling one thing from many others on which it depends simultaneously, with such skill that to do so requires no greater intelligence than to perform the most simple inference. Questions of this kind, since they are mostly abstract and occur almost exclusively in arithmetic and geometry, will seem to be of little use to those who are inexpert. But I advise those who wish to understand perfectly the latter part of my method, in which we discuss all other questions, that they should spend a long time in learning this method and practising it.

RULE THIRTEEN

If we understand a question perfectly, it must be abstracted from every superfluous concept, reduced to its most simple form and divided by enumeration into the smallest parts possible.

This is the only respect in which we resemble dialecticians. Just as, in teaching the forms of syllogisms, they presuppose that their terms or

RULE THIRTEEN

the subject matter is known, so likewise we require in advance here that a question has been understood perfectly. But we do not distinguish, as they do, two extreme terms and a middle term,[31] but we consider the whole matter as follows. First, in every question it is necessary that something be unknown, because otherwise there would be no point in looking for it. Secondly, whatever is unknown must be designated in some way or other; otherwise we would not be directed to investigate it, rather than something else. Thirdly, it can be designated thus only by means of something that is known. All these things apply even in the case of imperfect questions. Thus if someone asks what is the nature of a magnet, the meaning of the terms 'magnet' and 'nature' is already known, and we are directed to ask about that rather than about anything else, and so on. But in order for the question to be perfect, we would also need it to be completely determined, so that nothing else is sought apart from what can be deduced from what is given. Thus someone may ask me what should be deduced about the nature of a magnet simply from the experiments or observations that Gilbert claimed to have made, apart from whether they are true or false.[32] Likewise, if someone asks what I judge about the nature of sound simply from the fact that three strings, A, B and C, emit the same sound and, among them, B is twice as thick as A but is not longer than it, and is stretched by a weight which is twice as heavy, while C is not thicker than A but is merely twice as long and is stretched by a weight that is four times heavier, etc. From these data it is easily understood how all imperfect questions can be reduced to perfect questions, and that will be explained at greater length in its appropriate place.[33] It is also clear how this rule may be implemented, so that a difficulty that is well understood may be abstracted from every superfluous concept and reduced in such a way that we would no longer think we are dealing with a particular subject matter, but that we are merely comparing, in general, certain magnitudes with one another. For example, once we have decided to consider only such and such experiences about the magnet, there is no further difficulty in abstracting our mind from all others.

It it also the case that the difficulty should be reduced to its simplest form, in keeping with Rules Five and Six, and that it should be divided in accordance with Rule Seven. For example, if I investigate the magnet on the basis of a number of experiences, I would go through them separately, one after another. Likewise, for sound, as already indicated, I would compare separately the strings A and B, and then A and C, and so on, and subsequently I would include all of them together in an adequate enumeration. These are the only three things that must be observed by the pure intellect concerning the terms of some proposition, before we progress towards its final solution, if it needs the following eleven rules. How it should be done will be clearer from the third part of this treatise. By 'questions', however, we understand everything in which truth or falsehood occurs; the various kinds of question must be listed to determine what we are able to offer about each one.

We have said already that falsehood cannot occur in the mere intuition of things, whether they are simple or composite. Such intuitions are not called 'questions' in this sense of the term, but they acquire that name as soon as we deliberate to make a definite judgement about them. For we do not classify as questions only the requests that others make of us; but the ignorance or, preferably, the doubt of Socrates was also a question, when Socrates first turned towards it and began to ask if it were true that he doubted about everything and then claimed that it was true.

For we are searching for things from words, or causes from effects, or effects from causes, or a whole or other parts from parts, or finally many of these things at the same time.[34]

We are said to look for things in words, whenever a difficulty consists in the obscurity of a discourse. This includes not only all riddles, such as the riddle about an animal raised by the sphinx which initially was four-legged, then had two legs and eventually was three-legged; similarly, the riddle of the fishermen, who are standing on the shore, equipped with fishing rods and hooks for catching fish, and who claim that they no longer have the fish that they had caught but instead that they have those that they have not so far been able to catch, etc. But it is also the case that most of those about which the learned dispute are almost always questions about words. We should not have such a low opinion of great minds, that we think they understand things badly whenever they do not express them in appropriate words. For example, when they apply the term 'place' to 'the surface of the surrounding

RULE THIRTEEN

body', they do not really conceive of some false thing; they merely abuse the word 'place', which, according to common usage, means that simple nature that is known in itself and in virtue of which something is said to be here or there. This consists completely in some relationship between the thing that is said to be in a place and the parts of external space; some people mistakenly called it 'internal place', when they noticed that the word 'place' was used to refer to the surrounding surface; and so on for other similar cases. These questions about words occur so often that, if there were always agreement among philosophers about the meaning of words, almost all their disputes would be resolved.

We look for causes from effects whenever we ask, about something, if it exists or what it is [. . .][35]

For the rest, when we are asked to solve some question, we often do not recognize immediately what type of question it is and whether we are looking for a thing [beginning] from words, or causes from effects, etc. Therefore it would seem to me to be completely useless to say more about these in particular. It would be briefer and more convenient if we pursue in order, and simultaneously, everything that needs to be done to resolve a question. And then, in any given question, we should above all else strive to understand distinctly what is being sought.

For some people frequently rush into investigating questions in such a way that they apply an aimless intelligence to their solution, before noticing by what signs they might recognize what is sought if they happened to find it. They are just as inept as a servant who is sent somewhere by their employer, and who is so keen to obey that they hasten to run off without getting instructions and without knowing where they were told to go.

But in every question, although something must be unknown – otherwise, it would be redundant to look for it – it must nevertheless be designated by conditions that are certain, so that we are directed to investigate one particular thing rather than another. But these are the conditions that, from the very beginning, we said must be investigated. This will be done if we turn the eye of our mind to intuiting them distinctly, one by one, inquiring diligently about the extent to which the unknown that we are looking for is limited by each of them. For human intelligence tends to go wrong in two ways here, either by assuming something more than what it was given in order to determine the question or, on the other hand, by omitting something.

One must take care not to assume that we have more data than we do, or that the data are more precise than they are, especially in the case of riddles and other questions that are invented artificially to mislead our minds; but it also applies in other questions when, in order to resolve them, something seems to be assumed as certain of which we are convinced, not by a reason that is certain, but merely by a long-established opinion. For example, in the riddle of the sphinx, it should not be thought that the word 'foot' means only the real feet of animals, but one should ask whether the word could be transferred to something else, as happens when it applies to a child's hands or to an old person's walking stick, because both of these are used like feet for walking. Likewise, in the riddle of the fishermen, we should take care lest the thought of fish would so preoccupy our mind that it would turn away from the thought of those animals that the poor often carry about with them unwillingly, and once they are caught, they throw them away. Likewise if one asks how the vessel was constructed such as the one we once saw; there was a column in the centre of it, on top of which there was an image of Tantalus, as if he were longing for a drink. The water poured into this vessel was perfectly contained in it as long as it was not high enough to reach the mouth of Tantalus. But as soon as it reached his unfortunate lips, it all flowed out immediately. It seems initially as if all the skill were in constructing this image of Tantalus, whereas, in fact, that in no way determines the question and is a mere supplementary feature. The whole difficulty consists in this: that we are asking how the vessel should be constructed so that all the water flows out of it as soon as it reaches a given height, but not before then. Likewise, given all the observations we have about the stars, if someone asks: 'What can we claim about their motions?' it should not be assumed gratuitously that the earth is immobile and is placed at the centre of things, as the ancients assumed, because it has seemed to us to be that way since our infancy. But this should be called into doubt too, so that we subsequently examine what we can judge for certain about this question. And likewise for other cases.

RULE THIRTEEN

We sin by omission, however, whenever we do not reflect on some condition that is required for the determination of some question, and is either explicitly stated in the question itself or is in some way implicit in it. For example, if one asks about perpetual motion – not the natural kind that occurs in the stars and in fountains, but that which results from human effort – and if someone thinks that they could create perpetual motion (as many have thought possible, believing that the earth is moved in a perpetual motion around its axis and that a magnet has all the same powers as the earth), by arranging a magnet so that it moves in a circle or communicates its motion and all its other powers to some piece of iron; if this happened, they would not have invented perpetual motion artificially, but would merely have used a natural perpetual motion in the same way as someone who arranges a wheel, in the current of a river, so that it moves constantly. That would therefore omit a condition which is required to determine the question, etc.

Once a question is adequately understood, one needs to see exactly where the difficulty lies, so that it may be more easily solved when it is abstracted from everything else.

Understanding a question is not always enough to know where its difficulty lies. It is also necessary to reflect on each of the things that are required in it, so that we may omit some of them if it seems that they are easy for us to discover; once they have been left aside, the thing that we are looking for will be all that remains. For example, in the question about the vessel described a short time ago, we shall easily discover how the vessel should be made. A column must be built in its centre, a bird must be painted on it, etc.; once all these have been set aside, as irrelevant to what needs to be done, the difficulty becomes clear: the fact that the water contained in the vessel all flows out, once the water level reaches a certain height. And the question is: how does that happen?

We say here, therefore, that the only thing worth the effort involved is to review, in an orderly way, all the things that are given in the question itself, and to reject those that we recognize as obviously irrelevant, to retain those that are necessary and to refer those that are doubtful to a more detailed examination.

RULE FOURTEEN

This should also be applied to the real extension of bodies and presented to the imagination in its entirety by means of bare figures. In that way it will be perceived much more distinctly by the intellect.

In order to use the assistance of the imagination also, it should be noted that whenever something unknown is deduced from something else that was previously known, it is not the case that some new kind of thing is thereby discovered; rather, one merely extends all that knowledge to the point where we perceive that the thing we are looking for shares in one way or another the nature of those things that were given in the question itself. For example, if someone is blind from birth, it should not be hoped that, by any reasoning process, we could ever make them perceive the true ideas of colours, in the same way that we have them when they are derived from the senses. But if someone has at least seen the primary colours at some stage, but has never seen the intermediate or mixed colours, it is possible for them to imagine the images of those they have not seen by a kind of deduction from their similarity with the others.[36] In the same way, if there is some kind of thing in the nature of the magnet such that our intellect has never perceived anything similar to it, it cannot be expected that we would ever get to know it by reasoning; we would have to be taught either by some new sensation or by the divine mind. But if we perceive distinctly the combination of already known things or natures which produce the effects that appear in the magnet, we shall consider that we have learned whatever can be provided in this context by human intelligence.

And indeed all the things that are already known, such as extension, shape, motion and similar things (which need not be listed here), are known by means of the same idea in different subjects; if a crown is made of silver, we do not imagine that its shape is different from that of a crown made of gold. This common idea is transferred from one subject to another only by means of a simple comparison, by which we assert that what is sought is in some respect similar to, identical with or equal to something given. Thus in every reasoning process, we know the truth exactly only by means of a comparison. For example, in the following case: all A is B, all B is C, therefore all A is C, what is sought and what is given – namely, A and C – are compared with respect to both being B, etc. But because, as we have already often warned, the forms of syllogisms are of no assistance in perceiving the truth of things, it will benefit the reader to understand that, having rejected them completely, absolutely all knowledge that is not acquired by means of a simple and pure intuition of a solitary thing, is acquired by comparing two or more things. Indeed, almost all the work of human reason consists in preparing for this operation; for when it is clear and simple, we need no assistance from any art, but only from the light of human nature, to intuit the truth which is reached by this operation.

It should be noted that comparisons are said to be simple and clear only when what is sought and what is given share equally in some nature; all other comparisons require a preparation, only because the common nature that is present in both of them is not present equally but in other proportions or ratios. Most human effort is devoted simply to reducing those proportions, so that an equality between what is sought and what is known is seen clearly.

It should also be noted that only what admits of degrees, and everything that is included in the term 'magnitude', can be reduced to such an equality. Thus when the terms of some difficulty have been abstracted from every subject in accordance with the preceding rule, we understand that we are dealing only with magnitude in general.

Thus if we are also to imagine something and not use pure intellect, except with the assistance of the images depicted in the phantasy, it should be noted, finally, that anything that can be said about magnitude in general can also be attributed to any particular magnitude.

It is easy to conclude from what has been said that it would be just as beneficial if we transfer what we understand can be said of magnitudes in general, to the species of magnitude that is depicted most easily and distinctly in our imagination; that this is the real extension of a body, abstracted from everything else apart from the fact that it has a shape, follows from what was said in Rule Twelve, where we conceived of the phantasy itself, together with the ideas it contains, as nothing but a real body that is really extended and has a shape.

This is also self-evident, since the differences between proportions are not shown more distinctly in any other subject; although one thing can be said to be more or less white than another, and one sound can be more or less sharp than another, etc., we cannot define precisely whether the excess in question is in the ratio of 2 to 1 or 3 to 1, etc., except by some analogy with the extension of a body with shape. Let it remain decided and fixed, therefore, that questions that are perfectly determined contain hardly any difficulty, except the difficulty of expressing proportions in the form of equalities; and that everything in which this precise difficulty is encountered can easily be separated, and ought to be separated, from every other subject and then transferred to extension and shapes. Therefore we shall discuss these exclusively from here on, up to Rule Twenty-five, and omit every other consideration.

We would prefer, at this stage, to encounter a reader who is favourably disposed to study arithmetic and geometry, but would prefer someone who is not familiar with those disciplines at all rather than someone who has been taught them in the usual way. For the use of the rules that I am offering here is much easier in learning those subjects – and is sufficient for that purpose – than in any other type of question. This is so useful in searching for a deeper wisdom, that I do not hesitate to say that this part of our method was invented not for mathematical problems but, rather, that one should hardly learn mathematics except in order to acquire this method. And I assume nothing from those disciplines except perhaps some things that are known in themselves and are obvious to everyone. But the knowledge of those disciplines which others usually have, even if it were not compromised by any obvious mistakes, is obscured by many misguided and poorly conceived principles, and I shall try to correct that, as we go along, in the following pages.

By 'extension' we understand everything that has length, breadth and depth, leaving aside the question whether it is a real body or merely a space. Nor does it seem to need any more explanation, since there is absolutely nothing that is perceived more easily by our imagination. However, since the learned often use such sharp distinctions that they dissipate the natural light of reason and discover obscurities in things that even peasants know, they should be warned that 'extension' does not refer to some distinct thing that is separate from the subject itself, and that we do not generally acknowledge those kinds of philosophical entity which do not really fall within the scope of the imagination. For even if someone can convince themselves, for example, that if something that is extended in nature were reduced to nothing it would not involve a contradiction for its extension to exist on its own, they would not use a physical idea to conceive this, but only the intellect while it is judging poorly. They would admit that themselves if they reflected carefully on the image of extension that they attempt to depict at that time in their phantasy. For they would notice that they do not perceive it as deprived of every subject but that the way in which it is imagined is completely different from what they think. Thus, whatever the intellect believes about the truth of the matter, those abstract entities are never formed in the phantasy in such a way that they are separated from their subjects.

Since we shall undertake nothing from here on without the help of the imagination, it is worth distinguishing carefully the ideas by which the particular meanings of words are to be proposed to our intellect. Therefore, we propose to consider the following three ways of speaking: 'extension occupies a place', 'a body has extension' and 'extension is not a body'.

The first of these shows how extension is understood as something extended. For I conceive of the same thing if I say: 'extension occupies a place', and if I say 'an extended thing occupies a place'. Thus, to avoid ambiguity, there is no advantage in using the term 'extended thing', for it would not express as distinctly what we are thinking about, namely, that a subject occupies some place because it is extended. And someone could understand 'what is extended is a subject occupying a place' as if I said: 'whatever is animate occupies a place'. That explains why we said that we would be concerned here with extension, rather than with an extended thing, even though we think that it should not be conceived otherwise than as something that is extended.

We now pass on to the phrase: 'a body has extension', in which we understand the term 'extension' as meaning something other than a body. But we do not form two distinct ideas in our phantasy – one of a body and the other of extension – but only one idea, that of an extended body. As far as the reality is concerned, there is no difference between saying 'a body is extended' and 'something extended is extended'. This is characteristic of things that exist only in something else and can never be conceived without a subject. It is different in the case of things that are really distinct from their subjects. If I were to say, for example, that 'Peter has wealth', the idea of Peter is clearly distinct from that of his wealth. Likewise, if I said that 'Paul is wealthy', I would imagine something completely different than if I said 'a wealthy person is wealthy'. By not recognizing this difference, many people falsely believe that extension includes something which is really distinct from that which is extended, in the same way that Paul's wealth is distinct from Paul.

Finally, if someone says: 'extension is not a body', the term 'extension' is understood very differently than it is above; and in this sense there is no characteristic idea in the imagination that corresponds to it, but this whole proposition is put together by the pure intellect which alone has the capacity to separate such abstract entities. This is an occasion for error by many people, who fail to recognize that extension understood in this way cannot be comprehended by the imagination, and who represent it to themselves as a true idea. Since such an idea necessarily includes the concept of a body, if they say that extension understood in this way is not a body, they are foolishly involved in saying that 'the same thing is simultaneously a body and not a body'. But it is very important to distinguish expressions in which terms such as: 'extension', 'shape', 'number', 'surface', 'line', 'point', 'unity', etc., have such a narrow meaning that they exclude something from which they are not really distinct, as when someone says: 'extension or shape is not a body', 'number is not a numbered thing', 'a surface is the limit of a body', 'a line is the limit of a surface', 'a point is the limit of a line', 'unity is not a quantity', etc. All these and similar propositions should be completely removed from the imagination if they are to be true, and for that reason we shall not be discussing them in what follows.

It should be carefully noted that, in all other propositions in which these terms retain the same meaning and are used in the same way, abstracted from their subjects, but not excluding or negating anything from which they are not really distinct, we can and ought to use the assistance of the imagination. For in that case even though the intellect attends precisely only to what a word means, the imagination ought to form a true idea of the thing, so that the same intellect can turn to other conditions that are not expressed by the word, if the need arises, and without ever judging foolishly that they have been excluded. Thus if there is a question involving number, we would imagine some subject that is measurable by many unities; and although the intellect may well reflect only on the multitude of these in the present case, we will take care subsequently that it not draw any conclusion in which we assume that the numbered thing is excluded from our concept. Those who attribute wonderful mysteries and sheer nonsense to numbers do this, and they would certainly not believe so much about them if they had not conceived of number as distinct from numbered things. Likewise, if we are discussing shape, we will think of ourselves as dealing with an extended subject – but merely from the particular perspective that it has a shape. And if we are discussing a body, we will think of ourselves as dealing with the same subject in so far as it has length, breadth and depth. If we are discussing surface, we think of the same thing as long and wide, omitting its depth but not denying it; if we are discussing a line, we think of it merely as long; if a point, the same, having omitted everything else apart from the fact that it is an entity.

Although I deduce all these things at length here, human intelligence is so prejudiced that I still fear that hardly anyone is completely free from the danger of losing their way, and that they will find any explanation of my meaning too brief, even in a lengthy discourse. Even the disciplines of arithmetic and geometry, although they are the most certain of all, may still mislead us here. For what arithmetician does not think that their numbers should not only be abstracted from every subject by the intellect, but that they should also be distinguished by the imagination. What geometer does not confuse the clarity of their object by inconsistent principles, when they judge that lines lack width and a surface lacks depth, and subsequently construct one from the other not recognizing that the line – from the motion of which they conceive that a surface is formed – is really a body; however, that which lacks breadth is nothing more than a mode of a body, etc. But in order not to delay too long in listing these things, it will be quicker to explain how we think our object should be conceived, so that we demonstrate as easily as possible whatever truth is to be found in arithmetic and geometry.

Here, then, we are concerned with an extended object, and we are considering absolutely nothing else in it apart from its extension, purposely avoiding the term 'quantity' because some philosophers are so subtle that they have distinguished quantity from extension; but we are assuming that all questions have been deduced to a point where the only thing we are looking for is to know a particular extension by comparing it with some other extension that is already known. Since we are not expecting knowledge of some new entity here, but wish merely to reduce the proportions – however complicated they may be – so that what is unknown is found to be equal to something known. It is true that all the differences in proportions that are found in other subjects can also be discovered between two or more extensions. Therefore it is enough for our project if we consider in extension itself all the things that may help us to explain differences in proportion, and there are only three of these, viz. dimension, unity and shape.

We understand dimension as merely the mode or respect in which something is considered to be measurable. Thus length, breadth and depth are not the only dimensions of a body, but weight is also a dimension according to which objects are weighed, and speed is a dimension of motion, and there are innumerable others like this. For division itself into many equal parts – whether it is a real division or merely a notional one – is, strictly speaking, the dimension according to which we number things; and the mode which gives rise to number is properly said to be a kind of dimension, although there is some difference in the meaning of the term. For if we consider the parts in relation to the whole, we are said in that case to count; if on the other hand we look at the whole in so far as it is distributed into parts, we measure it. For example, we measure centuries in years, days, hours and minutes; if, however, we count the minutes, hours, days and years, we add up eventually to centuries.

It is clear from this that there can be innumerably different dimensions in the same thing, and that they add nothing extra to the measured things, but that they are understood in the same way whether they have a real basis in the things themselves or have been thought up arbitrarily by our mind. For there is something real in the weight of a body, the speed of a motion or the division of a century into years and days; but this is not the case in the division of days into hours and minutes. However, they all function in the same way if they are considered simply as dimensions – which is what must be done here and in mathematical disciplines; it is the responsibility of physicists to investigate if they have a basis in reality.

The recognition of this fact throws a considerable light on geometry, because in geometry almost everyone mistakenly thinks that there are three types of quantity: the line, the surface and the body. It has already been remarked above, that a line or a surface is not conceived as really distinct from a body or from each other; for if they are considered simply, as having been abstracted by the intellect, then they are not different types of quantity any more than animal and living creature are distinct kinds of substance in the case of human beings. It should be noted, in passing, that the three dimensions of bodies, length, width and depth, are only nominally distinct; for in the case of some given solid, there is nothing to prevent us from picking any extension we wish as the length, another as the width, etc. And although these three dimensions have a real basis at least in every extended thing, when considered simply as an extended thing, we are no more concerned with them here than with innumerable others that are either constructed by the intellect or have some other basis in things. For example, if we wish to measure a triangle exactly, there are three things to be known about the thing itself, namely, the three sides, or two sides and one angle, or two angles and its area, etc. Likewise in the case of a trapezium, there are five things to be known; there are six in the case of a tetrahedron, etc. These may all be called dimensions. But if, here, we are to choose those that would help our imagination most, we shall not consider more than one or two of them as they are represented simultaneously in our phantasy, even though we understand that, in the problem we are concerned with, there are many others; for it is the role of a method to distinguish them into as many as possible, so that we consider only very few at one time but, none the less, consider all of them in succession.

Unity is the common nature that we said above ought to be shared equally by all the things that are compared with each other. Unless some unity is already determined in a problem, we may assume as a unit either one of the magnitudes already given, or any other magnitude, and this will be the common measure for all others. We shall also understand that there are as many dimensions in it as there are in the extremes that are to be compared, and we shall conceive of it either simply as some extended thing – abstracting it from everything else, and then it will be identical with the geometers' point, when they compose a line from its motion – or as some line, or as a square.

As regards shapes, it has already been shown above how the ideas of all things can be formed by means of these alone. It remains to point out at this stage that, from among the innumerable different kinds of shape, here we should use only those by which all the different ratios or proportions are most easily expressed. There are only two kinds of things that are comparable – multitudes and magnitudes[37] – and we also have two kinds of shapes for presenting them to our conception; for example, the points by which a triangular number is represented,

or a tree which explains someone's ancestry,

are shapes for representing multitudes; those, however, which are continuous and undivided, such as a triangle, square, etc.,

RULE FOURTEEN

represent magnitudes.

In order to explain which among all these figures we are going to use here, it should be realized that all the relations that may obtain between things of the same type are reducible to two kinds: viz. those of order and those of measure.

It should also be known that a significant effort is required to think of things in order, as can be seen throughout in this method, which teaches hardly anything else. There is no further difficulty in knowing an order, once it has been discovered; for we can easily review the individual parts that have been ordered by the mind, according to Rule Seven, because in this type of relation some parts are related to others in themselves and not by reference to a third term, as happens in the case of measurements, and for that reason we discuss the latter here. For I know the order between A and B, without considering anything else apart from these two terms; but I do not know the relation of magnitude between two and three without considering a third term, namely, the unity which is the measure that is common to both of them.

It should also be known that continuous magnitudes can sometimes be completely reduced to a multitude with the help of a borrowed unity, and that this can always be done at least partially; and the multitude of units can subsequently be arranged in order, so that the difficulty involved in knowing a measure will eventually depend simply on inspecting an order, and the greatest benefit of the method is in this progression.

It should be noticed, finally, that among the dimensions of continuous magnitude, none can be more distinctly conceived than length and width, and that we should not consider more than two at the same time in any figure, when comparing two figures with each other. For if we have to compare more than two dimensions with each other, method dictates that we review them in succession and consider only two at a time.

From what has been said, it follows easily that there is just as much reason to abstract propositions from the figures that geometers deal with, as from any other subject matter, if there is a question about them. For this purpose nothing should be retained but rectilinear and rectangular surfaces or straight lines – which we also call figures because, as we said above, by means of them we imagine no less of a genuinely extended thing than by means of a surface. Finally, by means of these figures, sometimes continuous magnitudes are represented, sometimes multitudes or numbers. There is nothing simpler for explaining all different relations that can be found by human effort.

RULE FIFTEEN

It is also helpful in most cases to make drawings of these figures, and to display them to the external senses so that, in this way, we will more easily keep our thought attentive.

It is self-evident how they should be drawn so that the images of them will be formed more distinctly in our imagination when they are displayed to our eyes. For, in the first place, we shall represent unity in three ways, viz. by means of a square, □, if we consider it as having length and width, or by a line, ———, if we consider it merely as having length, and finally as a point, ., if we think of it merely as that from which a multitude is constructed. But however it is represented and conceived, we shall always understand that it is something extended in every way and capable of having innumerable dimensions. The same applies to the terms of a proposition; if we have to consider two of its different magnitudes at the same time, we shall display them to our eyes by means of a rectangle, the two sides of which will be the two magnitudes proposed; in this way, if they are incommensurable with unity,

```

                    ┌─────────────────────────┐
                    │                         │
                    │                         │
                    └─────────────────────────┘

or thus,            ●           ●           ●

                            ●           ●           ●

or thus
                    ┌───────┬───────┬───────┐
                    │       │       │       │
                    ├───────┼───────┼───────┤
                    │       │       │       │
                    └───────┴───────┴───────┘
```

if they are commensurable. Nor is anything else required unless there is a question about a multitude of units. Finally, if we consider only one of their magnitudes, we shall represent it either by a rectangle, one side of which is the proposed magnitude (where the other side is unity), as follows:

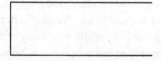

This is done whenever it is to be compared with some surface; or we shall represent it merely by means of a length, as follows, ———, if it is thought simply as an incommensurable length; or as follows, ..., if it is a multitude.

RULE SIXTEEN

In the case of things that may be necessary for the conclusion but do not need the mind's immediate attention, it is better to represent them by means of very compact symbols rather than by complete figures; for in this way memory could not go wrong, and meantime one's thought would not be distracted by retaining them while it is occupied in deducing other things.

For the rest, since we said that we should not contemplate more than two different dimensions in a single visual or mental intuition from among the innumerable dimensions that may be represented in our phantasy, it is worth while retaining all the others so that they easily come back whenever they are needed. Nature seems to have provided memory for this purpose. But since memory is often unreliable, and to avoid having to dedicate some of our mind's attention to repairing it while we are occupied with other thoughts, method has very appropriately invented the use of writing. Equipped with its help, we shall commit nothing further to memory here but, leaving the phantasy free and completely available to current ideas, we shall write down everything that has to be remembered. We shall do so with the briefest notes, so that once we have inspected each one distinctly, in keeping with Rule Nine, we shall be able (in accordance with Rule Eleven) to review everything by a very swift movement of thought and intuit simultaneously as many of them as we wish.

Therefore we shall represent by a single symbol (which can be formed in any way one wishes) anything that is to be regarded as a single item for solving a problem. But to make matters easy, we shall use the letters a, b, c, etc., for magnitudes that are already known, and A, B, C, etc., to express those that are unknown. We shall often prefix them with the numbers 1, 2, 3, 4, etc., to explain how many of them there are; and we shall also use numbers to indicate the number of the relations that should be understood in them. Thus if I write $2a^3$, that would be the same as if I said twice the number represented by the letter a, which contains three relations. In this way not only shall we save on many words, but – this is the principal thing – we shall display the terms of the problem so clearly and so simply that, while nothing useful is omitted, nothing superfluous will ever be included in them that would needlessly occupy the powers of our intelligence when many things have to be taken in at the same time.

In order to understand all these things more clearly, one should first realize that arithmeticians have usually represented individual magnitudes by many unities or by some number, but that we are abstracting here just as much from those numbers as we abstracted, a little while ago, from geometrical figures or anything else. We do this

RULE SIXTEEN

both to avoid the tedium of long and superfluous calculations, and especially so that the parts of the object that are relevant to the nature of the problem would always remain distinct and not become mixed up with useless numbers. Thus if someone is looking for the base of a right-angled triangle, the sides of which are given as 9 and 12, an arithmetician would say it is $\sqrt{225}$ or 15; but instead of 9 and 12 we would substitute *a* and *b*, and we would find the base is $\sqrt{a^2+b^2}$, and the two parts a^2 and b^2 which are confused together in the number would remain distinct.

It should also be noted that proportions which follow each other in a continuous sequence should be understood as a number of relations, something that others try to express in the familiar algebra through many dimensions and figures, the first of which they call the root, the second the square, the third the cube, the fourth the double-square, etc. I confess that I have been deceived by these terms for a long time. For it seemed that, after the line and the square, nothing clearer could be shown to my imagination than the cube and other figures that were similarly constructed. I also solved quite a number of problems with their help. But eventually, after much experience, I realized that, by using that way of thinking, I had never discovered anything that I could not have known much more easily and distinctly without it, and that such terms should be completely rejected lest they disturb our conception because the same magnitude, even though it is called a cube or a double-square, should never be represented to the imagination (according to the preceding rule) otherwise than as a line or surface. It should therefore be noted, above all, that the root, square, cube, etc., are nothing other than magnitudes that are continuously proportional, and that they are always assumed to be preceded by the unity that was adopted, about which we have already spoken above. The first proportional is referred to this unity immediately and by a unique relation; the second one, however, is related through the first, and thus through two relations; the third through the first and the second, and through three relations, etc. Thus from here on we shall call the magnitude that is called a 'root' in algebra the 'first proportional'; we shall call that which is called a 'square' the 'second proportional', and so on for the others.

RULES FOR GUIDING ONE'S INTELLIGENCE

Finally, it should be noted that, although here we abstract the terms of a problem from certain numbers in order to examine its nature, it may often happen that it can be resolved more easily in the given numbers than if it is abstracted from them. This happens because of a double use of numbers that we have already mentioned above, because they sometimes express an order and sometimes a measure. Thus once we have tried to express a problem in general terms, it should be recalled to the numbers given to see if, by chance, they could provide us with a simpler solution. For example, once we have seen that the base of a right-angled triangle with sides a and b is $\sqrt{a^2+b^2}$, 81 should be substituted for a^2 and 144 for b^2 which, added together, give 225, the root of which, or the mean proportional between 1 and 225, is 15. From this we shall know that the base, 15, is commensurable with the sides 9 and 12, not generally from the fact that it happens to be the base of a right-angled triangle, one side of which is related to the other as 3 is to 4. Those of us who seek a clear and distinct knowledge of things distinguish all these things; but the arithmeticians, who are satisfied if the sum they seek occurs to them even if they do not recognize how the solution results from what was given, do not do so, although this is what 'science' means, in a strict sense of the term.

However, it should generally be observed that nothing should ever be committed to memory that requires one's constant attention, if we can write it down, lest the redundant recollection would distract some part of our intelligence from the knowledge of some present object. A list should be made in which we would write the terms of a problem as they were presented on the first occasion; then how they may be

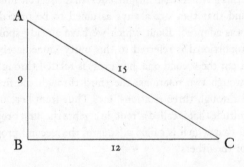

abstracted, and by means of what symbols they may be represented so that, once a solution has been found in those symbols, we can easily apply it, without any assistance from our memory, to the particular subject about which the problem arose. For nothing was ever abstracted except from something else less general. I shall write therefore as follows: we are looking for the base AC in the right-angled triangle ABC, and by abstracting the problem, so that in general what is sought is the size of the base from the length of the sides. Then for AB, which is 9, I put a; for BC, which is 12, I put b, and so on in other cases.

It should be noted that we will use these four rules further in the third part of this treatise, and that they will be understood more broadly than they have been explained here, as will be explained in the appropriate place.

RULE SEVENTEEN

Once a problem is posed, it should be reviewed directly by abstracting from the fact that some of its terms are known and others unknown, and by intuiting the mutual dependence of some terms on others by means of reliable reviews.

The preceding four rules have taught us how determinate and perfectly understood problems may be abstracted from specific subjects and deduced in such a way that nothing is subsequently sought, except that certain magnitudes are to be known from the fact that they are related to other given magnitudes through this or that relation. Now in the five following rules we shall explain how these same problems should be tackled so that, no matter how many unknown magnitudes are found in a given problem, they will all be subordinated to each other, and whatever the relation between the first and unity is, the second will be similarly related to the first, the third to the second, the fourth to the third, and so on for the others in sequence; thus no matter how many of them there are, they make up a sum equal to some known magnitude. We can do this by using a method which is so certain that we can safely claim that no amount of effort would be able to reduce them to simpler terms.

For present purposes, however, it should be noted that in every problem to be resolved through deduction there is some direct and easy approach by which we can pass most easily from some terms to others, and all the others are more difficult and indirect. To understand this, one should recall what was said in Rule Eleven, where we explained how, in a chain of propositions, if each one is linked with its neighbour, we easily perceive how even the first and last are reciprocally linked, although we do not deduce as easily the intermediate propositions from those at the extremity. However, if we intuit the dependence of individual propositions on one another without any interruption in the order, so that thereby we may infer how the last one depends on the first, we address the problem directly. On the contrary, if from the fact that we know that the first was connected to the last in a particular way, we wished to deduce the intermediate ones that linked them together, we would follow a completely indirect and opposite order. Since we are concerned here only with complex questions – i.e. those where some intermediate proposition is to be known, in a confused order, from the known extremes – all the skill of this project consists in the fact that, assuming the unknown are known, we are able to propose to ourselves an easy and direct method of discovery, even in the most complex problems. Nothing prevents this from always happening, since we have assumed from the beginning of this part that we knew that the relation between what is unknown in a given problem and what is known is such that the former is completely determined by the latter; thus if we reflect on those that occur to us first when we realize this determination and we are allowed to count the unknown among the known, so that, gradually, we can deduce from them by reliable reviews all the others that are also known, as if they were unknown, we shall be doing everything that is required by this rule. Examples of this, and also of most of the things we still have to discuss below, will be deferred until Rule Twenty-four, because they are explained more appropriately there.

RULE EIGHTEEN

Only four operations are required for this: addition, subtraction, multiplication and division; the last two of these should not be used often here, both to avoid needless complication and because they can be more easily done later.

A large number of rules often results from a lack of expertise in the teacher, and whatever can be reduced to one general rule is less clear if it is divided into many individual rules. For that reason we are reducing to only four headings here all the operations that should be used in reviewing questions, i.e. in deducing some magnitudes from others. How they manage to be sufficient will be discovered by explaining them.

If we arrive at knowledge of one magnitude from the fact that we know the parts of which it is constituted, that is done by addition. If we know a part from our knowledge of the whole and of the extent to which the whole is greater than the part, that is done by subtraction. As long as magnitudes are understood in an absolute sense, there is no other way of deducing one magnitude from others in which it is contained in some way or other. However, if one magnitude must be discovered from others which are completely different from it and in which it is in no way contained, it is necessary that it be related to them in some way. If this relation or connection is to be pursued directly, then one should use multiplication; if indirectly, one should use division.

To explain these two clearly one must realize that unity, about which we have spoken already, is the foundation and basis of all relations here, and that it occupies the first step in a series of magnitudes that are continuously proportional; that the given magnitudes are contained in the second step, while those which are sought are in the third, fourth and remaining places in the series, if the proportion is direct. However, if it is indirect, the magnitude that is sought is in the second and other intermediate steps, and what is given is in the final step.

For if it is said that unity is to a given a (or 5) as b (or 7) is to the magnitude that we are looking for, which is ab (or 35), then a and b are in the second place and ab (their product) is in the third. Likewise, if one also says that unity is to c (or 9) as ab (or 35) is to the number we are looking for, abc (or 315), then abc is in the fourth place, and results from two multiplications from ab and c, which are in the second place, and so on for other cases. In the same way, as unity is to a (or 5), likewise a (or 5) is to a^2 (or 25); and again, as unity is to a (or 5), so likewise is a^2 (or 25) to a^3 (or 125). Finally, as unity is to a (or 5), so a^3 (or 125) is to a^4 (which is 625). For whether a magnitude is multiplied by itself or by a completely different one, multiplication is the same in both cases.

If, however, we are told that unity is to a (or 5) – which is given as the divisor – as B (or 7), which is what we are looking for, is to ab (or 35) – which is the dividend – then the order is confused and indirect; as a result B, which we are looking for, remains unknown unless we divide ab (which is given) by a (which is also given). Likewise, if we are told that unity is to A (or 5) – which is what we are looking for – as A (or 5), which we are looking for, is to a^2 (or 25), which is given; or that unity is to A (or 5) which we are looking for, as A² (or 25), which we are looking for, is to a^3 (or 125), which is given; and so on for others. All these are included in division, although it should be noted that the latter cases are more difficult than the earlier ones, because the number we are looking for occurs more often there and therefore it involves more relations. The meaning of these examples is the same whether we say that the root is to be extracted from a^2 (or from 25) or the cube root from a^3 (or 125) and so on for the others; these are ways of speaking used by arithmeticians. Or if we are also to explain them in the language of geometers, it amounts to the same thing if one says that we need to find the mean proportional between the magnitude that is taken as given and is called unity, and that which is represented by a^2, or to find two mean proportionals between unity and a^3, and so on for others.

It is easy to conclude how these two operations are enough to find any magnitude that is to be deduced from others on the basis of some relationship. But once these things are understood, it follows that we must explain how these operations should be presented to the scrutiny of the imagination, and how they should be displayed even to the eyes, so that eventually we may explain their use or application.

If we are to do an addition or subtraction, we conceive of the object as a line, or as an extended magnitude in which length alone is to be considered. For if the lines a and b are to be added,

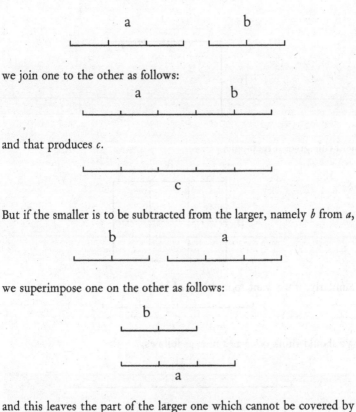

we join one to the other as follows:

and that produces *c*.

But if the smaller is to be subtracted from the larger, namely *b* from *a*,

we superimpose one on the other as follows:

and this leaves the part of the larger one which cannot be covered by the smaller one, namely:

In multiplication we also think of the given magnitudes as lines; but we imagine them forming a rectangle, for if we multiply *a* by *b*,

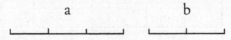

we put one line at right angles to the other as follows:

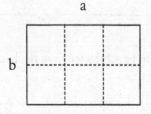

and this gives a rectangle:

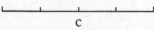

Similarly, if we want to multiply *ab* by *c*,

we should think of *ab* as a line, as follows:

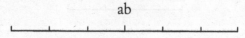

which gives the following for *abc*:

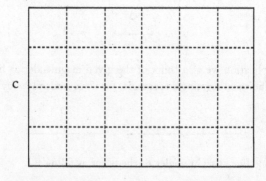

RULE EIGHTEEN

Finally, in the case of division where the divisor is given, we imagine the magnitude that is to be divided as a rectangle, one side of which is the divisor and the other the quotient. Thus if the rectangle *ab* is to be divided by *a*,

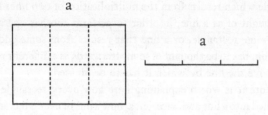

the width *a* is removed and that leaves *b* as the quotient:

On the other hand, if the same *ab* is divided by *b*, the height *b* is removed and the quotient would be *a*.

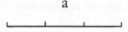

However, in divisions in which the divisor is not given but is merely represented by some relation – as when we are asked to extract the square root or the cube root, etc. – it should then be noted that the term to be divided and all the other terms should always be thought of as lines in a continuously proportional series, the first member of which is unity and the last of which is the magnitude to be divided. I shall explain in the appropriate place, however, how to find whatever mean proportionals one needs between the two ends of the series. It is enough to point out here that we are assuming we have not completed our discussions of these operations, since they have to be realized by means of indirect and reflective operations of the imagination, whereas we are currently discussing only questions that can be resolved directly.

As regards other operations, they can easily be performed in the way that we suggested thinking of them. It remains, however, to explain how their terms should be prepared. For even if we are free to think of its terms as lines or as rectangles when we first discuss some problem, and although we need not introduce any other figure (as was explained in Rule Fourteen), it often happens, none the less, that in the process a rectangle which results from the multiplication of two lines has soon to be thought of as a line, in order to perform another operation, or that the same rectangle, or a line that results from some addition or subtraction, has to be thought of soon afterwards as a different rectangle drawn above the line by which it has to be divided.

Therefore it is worth explaining here how every rectangle may be transformed into a line and, vice versa, how a line or even a rectangle may be transformed into a different rectangle, when one side of the latter is specified. This is very easy for geometers, once they realize that by 'lines' – whenever we compare them with some rectangle (as we are doing here) – we always understand rectangles with one side equal to the length that we have taken as unity. Thus this whole business is reducible to the following problem: given a rectangle, to construct another rectangle which is equal to it and of which one side is given.

I wanted to explain this lest I seem to omit anything, even though it is familiar even to neophytes in geometry.

RULE NINETEEN

Using this method of reasoning, we should look for as many magnitudes (expressed in two different ways) as there are unknown terms that we have assumed are known for the purpose of tackling a problem directly; for in that way we shall have the same number of comparisons between two equal terms.

RULE TWENTY

Once equations have been found, we should perform the operations that we have deferred, never using multiplication when division is appropriate.

RULE TWENTY-ONE

If there are a number of equations like this, they should all be reduced to one, i.e. to the equation the terms of which would occupy fewer places in the series of continuously proportional magnitudes, in accordance with which the terms should be arranged in order.

Notes

DISCOURSE ON METHOD

1. In French, *le bon sens*; as Descartes goes on to explain, '*le bon sens*' in this context means something like common sense, or the ability to use our innate intelligence to make sound judgements. The apparent equivalence of common sense and reason is evident from the phrase '*la raison, ou le sens*' (p. 6). Descartes also uses the phrase '*le sens commun*' in the *Discourse*, with two completely different meanings; one is synonymous with *le bon sens* (p. 10), while the other refers to a part of the brain (p. 40). I identify the latter by always using a definite article when referring to the common sense (see n. 22 below).
2. Descartes uses a scholastic distinction here, between the variable properties of things which are not part of their definition (such as the age or height of a human being), and the form or nature which includes all the properties that define something as a member of a particular species or class. According to the philosophers he invokes, one cannot be more or less a human being, although one may obviously be more or less old or tall.
3. The Jesuit college at La Flèche.
4. Breton is a distinct Celtic language that was widely spoken in Brittany in the seventeenth century, but was evidently regarded by some readers as inferior to French.
5. 'Our' suggests that Descartes had in mind the theology of the church to which he belonged, i.e. the Roman Catholic Church, rather than theology in general.
6. The so-called Thirty Years War, 1618–48.
7. Descartes attended the coronation of Ferdinand II in Frankfurt sometime between 20 July and 9 September 1619. According to his biographer, Baillet, the day referred to here was 10 November 1619, when he was returning to Duke Maximilian of Bavaria's army.
8. Ramón Lull (1232–1316), a Catalan philosopher, who invented a method for representing realities by letters and then combining the letters in various ways, in an art of combinations, to represent various possible worlds.

9. The four following rules represent a summary of the method that had been developed in greater detail in the unpublished *Rules* (see below).

10. For example, astronomy, physics, music, etc. Descartes himself uses the term 'mathematics' in this broad sense on some occasions, to include any scientific discipline which presupposes mathematical reasoning.

11. Refers to the doctrine of probabilism, which had been defended in theological controversies by the Jesuits; according to this view, it is morally right to act on the basis of merely probable opinions in circumstances in which we cannot be certain.

12. Descartes probably had in mind Seneca's *De vita beata* (*On a Happy Life*), which, together with the three maxims mentioned above, he discussed in a letter to Princess Elizabeth, 4 August 1645 (AT IV, 263).

13. Since the *Discourse* was a preface to a volume of scientific essays, this refers to problems that are analysed mathematically in the *Dioptrics* or the *Meteors*.

14. The Netherlands.

15. Refers to an early draft essay on metaphysical questions, from about 1628, which was unpublished when Descartes wrote the *Discourse*; it was eventually reworked and published as the *Meditations* (1641).

16. Moral certainty is the kind of certainty that we usually have about practical issues, and is enough to make decisions about actions even if, absolutely speaking, we may be mistaken. For example, we normally assume that a corked, labelled wine bottle contains wine, and we drink it without doing a chemical analysis of the bottle's contents.

17. i.e. *The World*, which Descartes decided not to publish following Galileo's condemnation by the Inquisition, because it contained a sympathetic discussion of the heliocentric theory.

18. In Chapter 7 of *The World*, Descartes constructs a hypothetical world which resembles the actual world, and he explains how the world descibed in his 'fable' could have evolved from the law-governed motions of an initial chaotic mass of matter created by God.

19. This suggests that the earth was created by God in the fully formed condition in which we observe it now; thus Descartes is not disagreeing with the account of creation found in the Book of Genesis.

20. William Harvey (1578–1657) published a book on the circulation of the blood, *De motu cordis*, in 1628. Descartes includes a marginal reference to 'Harvaeus, de motu cordis', at this point in the first edition.

21. In 1633, when *The World* was completed.

22. The common sense, in this context, is a part of the brain in which the sensory

23. An expression used by some Aristotelian philosophers to explain how matter can acquire new forms, which are said to be 'drawn from the potency of matter'.

24. Galileo's *Dialogue Concerning the Two Chief World Systems* (1632) was condemned by the Inquisition in Rome in 1633.

25. Descartes uses the French word '*expériences*' to mean observations, experiments, and experiences in a more general sense.

26. The contents include the *Dioptrics*, *Meteors* and *Geometry*.

27. A reference to the procedure adopted in the *Meditations* (1641), of publishing readers' objections and Descartes's replies to them as an integral part of the first edition. This suggestion was not implemented for the essays of 1637.

28. Refers to a machine for grinding lenses in Discourse 10 of the *Dioptrics*.

SELECTED CORRESPONDENCE, 1636–9

1. Marin Mersenne (1588–1648) was a Catholic priest, a member of the order of Minims and by far the most frequent among Descartes's correspondents.

2. The book was eventually published by Jan Maire at Leiden.

3. In the sentences omitted here, Descartes gives a brief summary of the contents of the *Dioptrics*, *Meteors* and *Geometry* in that order.

4. Constantijn Huygens (1596–1687) was a friend and supporter of Descartes, and father of the more famous Dutch physicist, Christiaan Huygens.

5. Jacob Golius [Gool], 1596–1667, who was professor of mathematics at Leiden University when Descartes studied there and was also a friend of Huygens.

6. Refers to the lengthy draft title proposed in the previous letter.

7. Mersenne and other correspondents had asked Descartes repeatedly to reconsider his decision not to publish *The World*; Mersenne had suggested publishing it with the *Discourse on Method* as an introduction.

8. See n. 15 for *Discourse on Method* above. Descartes was willing to discuss scepticism only when writing in Latin, so that it would not be accessible to untutored readers; when the Latin edition of the *Discourse* was published in 1644, there was no need to implement the suggestion in the text, because the topic of scepticism had been adequately discussed, in the mean time, in the *Meditations* (1641).

9. See above, p. 27.

10. The royal permission to publish from the court in Paris.

11. p. 53 above.

12. Here 'philosophy' means natural philosophy or physics.

13. Father Etienne Noël (1581–1660) had been Descartes's professor of philosophy at the college of La Flèche; Descartes had sent him a copy of the *Discourse*, inviting comments.

14. Father Noël had become rector of the college of La Flèche in 1636; hence his referring the *Discourse* to other Jesuits for their comments or evaluation.

15. Jean-Baptiste Morin (1583–1656) was Professor Royal of Mathematics at the Collège de France from 1629 until his death. He supported the theory, mentioned in this letter, that the earth was stationary.

16. p. 53 above.

17. On page 3 of the first edition, Descartes had written: 'Since I have occasion here only to speak about light in order to explain how its rays enter the eye and how they can be bent by various bodies that they encounter, there is no need for me to try to say what its real nature is. I believe it will be enough if I use two or three analogies that help to conceive of light in the way that seems most appropriate to me ... In doing this I imitate astronomers who, although their hypotheses are almost all false or uncertain, still manage to draw from them many consequences which are very true and very certain, because they correspond with various observations that they have made.' (AT VI, 83.)

18. Morin's book was called: *Letters written to Mr Morin by the most celebrated astronomers of France, approving his invention of longitudes, etc.* (Paris: Morin and Libert, 1635).

19. p. 48 above.

20. p. 52 above.

21. Antoine Vatier (1596–1659) was a Jesuit at La Flèche.

22. 18 October.

23. This is not meant to imply that women were less intelligent than male readers, but merely that the book was written in the vernacular and was therefore accessible to those who had not studied philosophy (which was taught only in Latin).

24. The unpublished *World, or a Treatise on Light*.

25. On Descartes's account, the substance of bread is changed during the Mass, as was taught by the Council of Trent. But the appearances of bread and wine that remain unchanged are not real qualities of a substance which is no longer present, but are merely subjective states of the perceiver that are caused by God.

26. The order in question is the Jesuits, and the school was La Flèche, which Descartes had attended.

27. Henri Regnier (1593–1639) was appointed the first professor of philosophy at the newly established University of Utrecht in 1634, and helped to introduce Cartesian philosophy to the curriculum there.

28. p. 19, line 3, above.

29. p. 20, line 17, above.

30. p. 19, line 5, above.

31. p. 22, line 6, above.

32. The identification of a virtue as a mean between two contrasting vices was a characteristic scholastic approach to their definition.

33. p. 27, line 3, above.

34. Sceptics named after the fourth-century BC sceptic, Pyrrho of Elis, who recommended that we not make judgements about anything because we lack certainty.

35. pp. 40–41 above.

36. See the *Meteors* (AT II, 233), where Descartes compares the parts of water with eels.

37. Witelo (*c*.1230-*c*.1275), a Polish natural philosopher whose refraction tables were apparently used by Descartes.

38. Pierre de Fermat (1601–55), one of the foremost mathematicians of the seventeenth century.

39. See n. 17 above.

40. Here Descartes does not cite the original letter exactly.

41. *Discourse on Method*, p. 53 above.

42. Typical scholastic hypotheses, used in the explanation of natural phenomena, appealed to entities such as these.

43. Florimond Debeaune (1601–52) was a French mathematician who wrote a short commentary on Descartes's *Geometry*.

44. Refers to *On Truth, insofar as it is distinct from Revelation, Probability, Possibility, and Falsehood*, by Edward Herbert, Lord Cherbury, 1637. The first two editions were published in Latin as *De Veritate* in 1624 and 1633.

45. Johannes Bannius (1579–1644), a student of musical theory and Catholic Archbishop of Harlem.

46. I. Boillau, *Philolaus, or a Dissertation on the True System of the World, in Four Books* (Amsterdam, 1639).

47. Lord Cherbury's book on truth (see n. 44 above).

48. A well-known scholastic definition of motion.

THE WORLD

1. Descartes relies on a distinction commonly made at the time between natural signs of mental states (such as tears as a sign of sorrow) and conventional signs (such as the words by which we express our sorrow).

2. The word 'stars' is used here to include the sun, the stars and any other heavenly body which is a source of light.

3. Here Descartes refers to the theory of real qualities, forms and powers which was proposed by scholastic philosophers (n. 42 for *Selected Correspondence*, p. 197 above); he suggests that it be replaced by mechanical explanations of natural phenomena (such as the burning of wood).

4. The parts of a flame move so quickly that, despite being so small that they are unobservable, the product of size and speed gives them enough force to break apart a body such as a piece of wood.

5. See the *Dioptrics* (1637), Discourse II, where Descartes explains the sine law of refraction by analysing the motion of a ray of light into two components, the direction in which it moves and the speed of its motion.

6. Descartes is avoiding the implication that there is a vacuum in nature, or that there are empty gaps between the small parts of a fluid body.

7. See *Rules*, Rule Twelve, p. 161 below.

8. i.e. bodies located below the clouds.

9. According to Aristotle and the scholastic philosophers who adopted his theory of elements, the four basic elements were defined by combinations of wet/dry and hot/cold; earth was cold and dry, water was cold and wet, air was hot and wet, and fire was hot and dry.

10. In the scholastic language used here, an 'accident' is a non-essential quality of something.

11. Descartes quotes, in Latin, a standard scholastic definition of motion. He goes on to provide a French translation which is found in English in the text.

12. Earlier theories suggested that the air that is displaced, for example, by a moving arrow, turns behind the arrow and pushes it in its original path.

13. If one body strikes another, it can overcome the latter's resistance and thereby expend most of its own motion; or it could submit to the other body and, by obeying it, acquire some of the force which the dominant body expends in the collision.

14. The sling in question is a piece of leather, folded, and the two ends are tied to a string. The stone is then in the fold or middle of the leather.

15. Descartes provides such rules in the *Principles of Philosophy* (1644), Part II, arts. 45–62.

RULES FOR GUIDING ONE'S INTELLIGENCE

1. In Latin, *de bona mente*; this is the *bon sens* (the ability to make sound judgements) which features in the first sentence of the *Discourse on Method*, and which was translated in the first Latin edition of the *Discourse* as *bona mens*.

2. The Latin term is *scientia*, which meant systematic, well-founded knowledge rather than scientific knowledge in the modern sense of the term.

3. Literally, 'to withdraw our hands from the rod'.

4. Here Descartes adverts to the sensitive question of the extent to which the content of religious faith is amenable to rational argument. In so far as faith presupposes understanding, what we believe on faith can be clarified only by intuition or deduction. See n. 25 below.

5. Alexandrian mathematicians of the fourth and third centuries B C respectively; a Latin translation of the *Mathematical Collection*, by Pappus, became widely available in the late sixteenth century.

6. Descartes had in mind the classification of things as substance, and as various general types of quality or relation.

7. i.e. they form a geometrical progression.

8. In modern notation, find the value of x in the equation: $3/x = x/12$.

9. The Adam and Tannery edition has 'movement of the imagination' here, but I follow Crapulli and what is implied by other references to the same operation, and substitute 'movement of thought'.

10. The fifth rule is spelled out in more detail in Rule Six rather than Rule Five.

11. In other words, the problem to be solved.

12. 'By 'illumination' Descartes means the action by means of which luminous bodies affect others at a distance. He did not think that light was something like a body that actually moved from one place to another; rather it is merely a tendency to move, an action or a force which is transmitted through a medium without anything in the transparent medium actually moving at all.

13. See n. 1 above.

14. Here something is missing from the available MSS.

15. i.e. Rule Nine.

16. Only part of Book Two survives, and none of the projected Book Three.

17. Descartes refers to the bare (*nuda*) force being communicated; he means that the force is the only thing that is communicated, and that its communication from one body to another or between different parts of the same body does not involve the movement of a body.

18. The forms of the syllogism, which were accepted among scholastic philosophers as the best guide to logical reasoning.

19. Here Descartes assumes the widely used scholastic distinction between matter and form or, in this context, between structure and content. Syllogistic arguments are concerned only with logical form or structure; therefore, if there is truth in the conclusion of a syllogism, it must have been contained in the premisses (the matter), since the form or structure alone cannot generate truth.

20. In scholastic philosophy, the mind is a form, of which the corresponding matter is the body.

21. See n. 22 for *Discourse on Method* above (p. 194).

22. Descartes uses two words in Latin, *phantasia* and *imaginatio*, for this faculty. I have reflected his choice of terms by using 'phantasy' and 'imagination' in English.

23. Descartes refers to these above (p. 154) as 'shapes or ideas', so that the word 'shapes' here refers to the ideas that arise in the mind from images in the common sense.

24. That is, to the objects of knowledge, in contrast with the subject of knowledge just discussed.

25. Refers to conviction that results from religious faith, inspired by the superior power of God through revelation.

26. See below, the incomplete eighth item (p. 163).

27. Aristotle, *Physics*, IV, 4, 212^{a}5.

28. See Descartes's *Principles*, Part I, art. 10.

29. The previous paragraph was the fourth item on a list; but this follows from an earlier listing, which concluded with 'seventhly' (p. 161).

30. The text is incomplete here. The same issue is taken up again in Rule Thirteen (pp. 166–7).

31. The so-called major term and minor term are the two extreme terms of a syllogism and occur as the predicate and subject terms, respectively, of the conclusion. The middle term occurs in both premisses but not in the conclusion, and provides the logical link between the extreme terms.

32. William Gilbert (1540–1603) claimed, in *De magnete* (1600), to have made progress in understanding magnetism precisely because of well-conducted experiments and observations.

NOTES

33. This was to be done in the third group of twelve rules, which apparently were never written.

34. This repeats a sentence left incomplete above, where Descartes suggests that we need a method of discovery. This rule is translated loosely by Arnauld and Nicole in *Logic or the Art of Thinking*, ed. Jill V. Buroker (Cambridge University Press, 1996), Part IV, Chapter 2, pp. 234–7.

35. Here again the text is incomplete.

36. The Amsterdam edition has a marginal note here: 'this example is not completely true, but I had no better example to explain something that is true'. This question anticipates the so-called Molyneux problem subsequently discussed by Locke and other empiricists.

37. Refers to discontinuous and continuous magnitudes.

Index

abstraction, 165, 171, 182–3
Académie Royale des Sciences, xi
alchemists, 125
algebra, 15, 17, 18, 126, 128, 183
anaclastic, 139–40
analogy, 140, 154, 172
a posteriori, 67
a priori, 67, 112
Archimedes, 73, 74
arithmetic, 18, 120–21, 126, 127, 129, 164, 172, 175, 178
Aristotle, xv, xxi, xxiii, xxx, 49, 63, 122
Arnauld, Antoine, 115
arts, contrasted with science, 117
assumptions, 53, 67, 74, 75, 76–7
astrologers, 130
atoms, 92
automota, 40–41, 72–3

Bacon, Francis, xv
Baillet, Adrien, 115
Bannius, Johannes, 78
Beeckman, Isaac, xxix
Bellarmine, Robert Cardinal, xxi
blood circulation, 35–9
Boyle, Robert, xi, xxii, xxiii

Brahe, Tycho, xi
burning, 87–9, 92

Calvinists, 68
certainty
 mathematical, 121
 metaphysical, 28
 moral, 28
Clerselier, Claude, xiv, 115
clarity and distinctness, 25, 28, 144
common notions, 157
common sense, 193n1
 sound judgement, 5, 12, 20, 118, 140
 part of the brain, 40, 153, 154
Copernicus, Nicolas, xi, xii, xxi
Crapulli, Giovanni, 115
creation, 33, 42, 102–3

Debaune, Florimond, 77
deduction, 120, 123–4, 125, 149, 160, 163
demonstration, xv–xxii, 12, 16, 24, 33, 36, 43, 50, 60, 73–4, 75, 111–12, 121, 122, 163
dimension, 176
Diophantes, 128
direct/indirect questions, 135
discovery, method of, xx, xxix, xxxii, 65
dreams, 28–9
dualism, 71

205

INDEX

elements, 98–102
  air, 98
  earth, 98–9
  fire, 97–8
enumeration, 16, 18, 135–8, 143, 149
  adequate, 136–7, 141, 161, 166
  complete, 137, 151
  well-ordered, 137–8
experience (observation, experiment), 23, 45–6, 50–52, 120–21, 144, 159, 165
explanation, xvii, xxii–xxiii, 65
extension, definition of, 172–4
  and quantity, 176

fable, 7, 8, 31, 102, 112
faculties for knowing, 141, 143, 151, 154–5
faith (religious), 124, 160
Fermat, Pierre, xi, 74
forms, scholastic, xvii, xxiii, xxvi, 31, 77, 103
foundations, 13, 24

Galen, xv
Galilei, Galileo, xi, xii, xvi, xxi, xxx, 43
Gilbert, William, xi, 165
geocentric theory, 168
geometry, 15, 17, 120–21, 125, 126, 127, 129, 164, 172, 175, 188
God's existence, proof of, 26–7, 65
Golius, Jacob, 58

Harvey, William, xi, 36
heart, functioning of, 34–9
Herbert, Edward (Lord Cherbury), 78–9
Huygens, Constantijn, 58

Huygens, Christiaan, xi, xxii
hypotheses, xvii, xxviii, 63–4, 66, 76, 160

ideas, of sensations, 85–6, 89
imagination, 28, 59, 123, 154, 155, 160, 170, 175
induction, *see* enumeration
inertia, 77–8
intellect, 143, 154–5, 166
intuition, 123–4, 125, 130, 132, 135, 136–7, 144, 149, 161, 166, 167, 171
'I think, therefore I am', 25, 70–71, 123

Kepler, Johannes, xi
Kuhn, Thomas, xxiii

language, 40–42
laws of nature, xxvii–xxviii, 30, 31, 39, 105–12
La Flèche, xii, 7
Leibniz, Gottfried Wilhelm, 115
light, 74, 75, 85–9, 140
logic, 15, 121, 126, 148, 164–5, 171
Lull, Ramón, 15

magnetism, 163, 165, 170
Marion, Jean-Luc, 115, 116
mathematics, 9, 16–17, 23, 127, 129, 134, 139, 172
mathesis, xxix, 128, 129, 130, 139
matter, 93, 103–4
  essence of, 104
  prime matter, 103, 104
measure, 179
mechanism, xxv
medicine, 44, 54

memory, 136, 149, 150, 154, 155, 181–2
Mersenne, Marin, xviii, xxi, 57, 58, 73, 78, 79
metaphysics, xvii, 60, 67
method, scientific, xiv–xv, 61
　a priori, xxx, 67
　definition of, 125
　hypothetico-deductive, 75
　innate, 142
　rules of, 16
models, xvii, xxi
motion
　circular, 94–5
　definition of, 106–7, 162
　determination of, 88, 198n5
　force of, 88, 90–91
　laws of, *see* laws of nature
　quantity of, 77, 109
moral maxims, 19–22
　provisional, 69
Morin, Jean-Baptiste, xxi, xxvi, 62, 75

natural light (of reason), 11, 22, 79–80, 118, 125–6, 132, 157, 171, 172
natural power, 74, 140, 145
natural signs, 41–2, 72
nature, 105
Newton, Isaac, xi
Nicole, Pierre, 115
Noël, Etienne, 61

objects of knowledge
　absolute/relative, 131–3
　complex/simple, 131, 143, 156–9
occult powers, 162
order, 16, 18, 130–31, 179
Occam's principle, xxvi

Pappus, 128
perfect/imperfect problems, xxxi–xxxii, 164–5
Philolaus, 78
phantasy, 141, 154, 160, 171, 173, 177
philosophy, 9–10, 14
Plato, 122
Plempius, Vobiscus Fortunatus, xxvi
Poisson, Nicolas, 115
Port-Royal Logic, 115
prejudice, 16, 18, 73
probable, 20, 36, 119–20, 123
Ptolemy, xv, 73, 74
pyrrhonists, 71

qualities, 98–9, 104
　see forms

Ramus, Peter, xv
reason, definition of, 5, 6
Reneri (Regnier), Henri, 68
Royal Society, xi

scepticism, 22, 25
scholastic terminology, 26, 123
scholastic philosophers, 28, 120, 125, 131
science, meaning of, 3–4, 119, 143, 184
scientific societies, xi–xii
scientific revolution, xi
seeds of truth, 45, 126, 128
Seneca, Lucius Annaeus, 194n12
sensations, xvi–xvii, 24, 59, 85–7, 96, 152–4
Socrates, 166
soul
　human, 33–4, 42, 71, 137
　of animals, 42, 71

substance
  material, 137
  spiritual, 25

Thales, 67
theology, 9, 62, 78
thought, definition of, 69
transubstantiation, 68, 195*n*25
truth, 79

unobservable entities, xxvi–xxvii, xxxviii, 90

vacuum, 93–7, 103, 161
Vatier, Antoine, xix, xx, 64
vernacular language, 54

Wallace, William A., xxx
weight, 78
Witelo, 73, 74
Willis, Thomas, xi
women readers, 65

Zabarella, Jacopo, xv

DISCOURSE ON METHOD AND THE MEDITATIONS

René Descartes

'I think, therefore I am'

Widely regarded as the founder of modern western philosophy, René Descartes (1596–1650) sought to look beyond the established tenets of earlier thinkers and establish a philosophical system more appropriate for the scientific age in which he lived: one based on the power of reason. In the *Discourse*, he details the famous 'Method' by which he strove to develop this system – beginning with the concept of universal doubt and going on to deduce, from the celebrated phrase *cogito ergo sum*, both his own existence and that of God. The *Meditations* is a penetrating exploration of his beliefs, which considers in depth the nature of God, the human soul, and the concept of truth and error. Profoundly influential and original, both works have been the source of outrage and inspiration for generations of readers and philosophers as varied as Leibniz, Hume, Locke and Kant.

Translated with an introduction by F. E. Sutcliffe

ISBN: 978 0 14 044 206 9

LEVIATHAN

Thomas Hobbes

'The life of man, solitary, poore, nasty, brutish, and short'

Written during the chaos of the English Civil War, Thomas Hobbes' *Leviathan* asks how, in a world of violence and horror, can we stop ourselves from descending into anarchy? Hobbes' case for a 'common-wealth' under a powerful sovereign – or 'Leviathan' – to enforce security and the rule of law, shocked his contemporaries, and his book was publicly burnt for sedition the moment it was published. But his penetrating work of political philosophy opened up questions about the nature of statecraft and society that influenced governments across the world.

Edited with an Introduction by Christopher Brooke

AN ESSAY CONCERNING HUMAN UNDERSTANDING

John Locke

'It matters not what men's fancies are, 'tis the knowledge of
things that is only to be prized'

In *An Essay Concerning Human Understanding*, first published in 1690, John Locke (1632–1704) provides a complete account of how we acquire mathematical, natural, scientific, religious and ethical knowledge. Rejecting the theory that some knowledge is innate, Locke argues that it derives from sense perceptions and experience, as analysed and developed by reason. While defending these central claims with vigorous common sense, Locke offers many incidental – and highly influential – reflections on space and time, meaning, free will and personal identity. The result is a powerful, pioneering work, which, together with Descartes's works, largely set the agenda for modern philosophy.

Edited with an introduction and notes by Roger Woolhouse

A TREATISE OF HUMAN NATURE

David Hume

'Human Nature is the only science of man, and yet has hitherto been the most neglected'

One of the most significant works of Western philosophy, Hume's *Treatise* was published in 1739–40, before he was thirty years old. A pinnacle of English empiricism, it is a comprehensive attempt to apply scientific methods of observation to a study of human nature, and a vigorous attack upon the principles of traditional metaphysical thought. With masterly eloquence, Hume denies the immortality of the soul and the reality of space; considers the manner in which we form concepts of identity and cause and effect; and speculates upon the nature of freedom, virtue and emotion. Opposed both to metaphysics and to rationalism, Hume's philosophy of informed scepticism sees man not as a religious creation, nor as a machine, but as a creature dominated by sentiment, passion and appetite.

Edited with an introduction by Ernest C. Mossner

A PHILOSOPHICAL ENQUIRY INTO THE SUBLIME AND BEAUTIFUL

Edmund Burke

'The great has terror for its basis ... the beautiful is founded on mere positive pleasure'

Edmund Burke was one of the foremost philosophers of the eighteenth century and wrote widely on aesthetics, politics and society. In this landmark work, he propounds his theory that the sublime and the beautiful should be regarded as distinct and wholly separate states – the first an experience inspired by fear and awe, the second an expression of pleasure and serenity. Eloquent and profound, *A Philosophical Enquiry* is an involving account of our sensory, imaginative and judgemental processes and their relation to artistic appreciation. Burke's work was hugely influential on his contemporaries and also admired by later writers such as Matthew Arnold and William Wordsworth. This volume also contains several of his early political works on subjects including natural society, government and the American colonies, which illustrate his liberal, humane views.

Edited with an Introduction and notes by David Womersley

ISBN: 978 0 14 043 625 9

CANDIDE, OR OPTIMISM

Voltaire

'You have been to England ... Are they as mad as in France?'

Brought up in the household of a German Baron, Candide is an open-minded young man whose tutor, Pangloss, has instilled in him the belief that 'all is for the best'. But when his love for the Baron's rosy-cheeked daughter is discovered, Candide is cast out to make his own fortune. As he and his various companions roam over the world, an outrageous series of disasters befall them – earthquakes, syphillis, a brush with the Inquisition, murder – sorely testing the young hero's optimism. In *Candide*, Voltaire threw down an audacious challenge to the philosophical views of his time, to create one of the most glorious satires of the eighteenth century.

Translated by Theo Cuffe
with an Introduction by Michael Wood

THE SOCIAL CONTRACT

Jean-Jacques Rousseau

'Man was born free, and he is everywhere in chains'

These are the famous opening words of a treatise that has not ceased to stir debate since its publication in 1762. Rejecting the view that anyone has a natural right to wield authority over others, Rousseau argues instead for a pact, or 'social contract', that should exist between all the citizens of a state and that should be the source of sovereign power. From this fundamental premise, he goes on to consider issues of liberty and law, freedom and justice, arriving at a view of society that has seemed to some a blueprint for totalitarianism, to others a declaration of democratic principles.

Translated and Introduced by Maurice Cranston

THE NEED FOR ROOTS

Simone Weil

'Rootedness is perhaps the most important and least known human spiritual need'

What do humans require to be truly nourished? Simone Weil, one of the foremost philosophers of the last century, envisaged us all as being bound by unconditional, eternal obligations towards every other human being. In *The Need for Roots*, her most famous work, she argued that our greatest need was to be rooted: in a community, a place, a shared past and collective future hopes. Written for the Free French movement while she was exiled in London during the Second World War, Weil's visionary combination of philosophy, politics and mysticism is her answer to the question of what life without occupation – and oppression – might be.

Translated by Ros Schwartz
with an introduction by Kate Kirkpatrick

THE PRINCESSE DE CLÈVES

Madame de Lafayette

'Am I to be loved by the most adorable woman in the world ... only so that I may the more cruelly suffer the pain of her unkindness?'

Set towards the end of the reign of Henri II of France, *The Princesse de Clèves* (1678) tells of the unspoken, unrequited love between the fair, noble Mme de Clèves, who is married to a loyal and faithful man, and the Duc de Nemours, a handsome man most female courtiers find irresistible. Warned by her mother against admitting her passion, Mme de Clèves hides her feelings from her fellow courtiers, until she finally confesses to her husband – an act that brings tragic consequences for all. Described as France's first modern novel, *The Princesse de Clèves* is an exquisite and profound analysis of the human heart, and a moving depiction of the inseparability of love and anguish.

Translated with an Introduction and Notes by Robin Buss

LES MISÉRABLES

Victor Hugo

'They fought, hand to hand, for every inch of ground, using pistols, swords, fists, at a distance, at close quarters, from above, from below, from everywhere'

Les Misérables, or 'the wretched', is Victor Hugo's epic novel of redemption, sacrifice, love and suffering, set against the turbulent backdrop of early nineteenth-century France. It tells the story of ex-convict Jean Valjean, who has spent nineteen brutal years in chains after stealing a loaf of bread. Saved by an act of Christian charity, he is then offered a chance of happiness when he encounters the downtrodden Fantine and vows to rescue her daughter Cosette – but is constantly pursued by the implacable policeman Javert, who will not let him escape his past. An instant bestseller on publication in 1862, *Les Misérables* is a novel of intense emotional power, weaving together individual stories with real-life historical events to create a rich, imaginative drama of human life.

Translated with Notes by Christine Donougher
with an Introduction by Robert Tombs

ISBN: 978 0 24 124 874 4